T0202800

UNITEXT for Physics

UNITEXT for Physics series, formerly UNITEXT Collana di Fisica e Astronomia, publishes textbooks and monographs in Physics and Astronomy, mainly in English language, characterized of a didactic style and comprehensiveness. The books published in UNITEXT for Physics series are addressed to graduate and advanced graduate students, but also to scientists and researchers as important resources for their education, knowledge and teaching.

More information about this series at http://www.springer.com/series/13351

Ettore Vitali · Mario Motta
Davide Emilio Galli

Theory and Simulation of Random Phenomena

Mathematical Foundations and Physical Applications

 Springer

Ettore Vitali
Department of Physics
College of William and Mary
Williamsburg
USA

Davide Emilio Galli
Department of Physics
University of Milan
Milan
Italy

Mario Motta
Division of Chemistry and Chemical
  Engineering
California Institute of Technology
Pasadena
USA

ISSN 2198-7882   ISSN 2198-7890   (electronic)
UNITEXT for Physics
ISBN 978-3-030-08032-7   ISBN 978-3-319-90515-0   (eBook)
https://doi.org/10.1007/978-3-319-90515-0

Printed on acid-free paper

This Springer imprint is published by the registered company Springer International Publishing AG
part of Springer Nature
The registered company address is: Gewerbestrasse 11, 6330 Cham, Switzerland

To our families and friends

Contents

Introduction

The importance of probability theory in several fields of pure and applied science can hardly be overrated. In particular, it is a cornerstone in which several branches of physics and mathematics have their roots.

Probability theory is the theoretical framework underlying statistics, and thus enables to rigorously extract information from experimental data with enormous implications in all fields of science. The language of several branches of modern theoretical physics, from quantum mechanics to quantum field theory and statistical physics, is formulated in terms of random variables and processes. One could think that celebrated Gibbs ensembles of classical statistical mechanics, for example $Q^{-1} exp(-\beta H)$, are, in the classical case, nothing but probability densities. In quantum mechanics, the wave function of a physical system, the quantum state, say $\Psi(x)$, allows one to compute probability densities for any observable. Even quantum field theory, when formulated using euclidean Feynman path integrals, is a probability theory on the space of configurations of the quantum fields.

The ubiquitous presence of probability theory in modern science and the advent of more and more powerful computational resources have conferred a role of increasing relevance to computational methods in ductile and powerful techniques for investigating reality. Actually, computational physics lies somehow in the middle between theoretical and experimental physics: experiments are performed on a computer relying on theoretical models. One creates his/her own virtual laboratory and performs his/her own experiments inside it by sampling a random variable whose expected value is the solution of the physical problem under study. The mathematical foundation of most of the methodologies in computational physics lies, again, in probability theory. Just to give an example, the very famous Metropolis algorithm, which enabled the quantitative study of all Gibbs ensembles, ranging from classical fluids to biological molecules, has its roots in an advanced topic in probability theory, the theory of Markov chains.

Finally, the study of complex systems, with plenty of applications in biology, physics, chemistry, economy, computer science, and many other fields, has become during last decades an extremely attractive research field. Probability theory is naturally essential for the mathematical modeling, and simulation is a crucial tool to understand the behavior of such systems.

A robust background in probability theory is thus arguably a mandatory requisite for graduate students in physics or related disciplines. This book is meant to guide a student from the very foundations to more advanced topics, like stochastic processes and stochastic differential equations. This material is typically dealt with in advanced textbooks about mathematics, which, due to the volume and the high level of sophistication of the formalism, are sometimes hard to read for a reader with a non-strictly mathematical background. We present the material in such a way to provide a link between basic undergraduate level textbooks and such advanced books. In doing this, we will keep extreme attention to the mathematical rigor and, contemporarily, we add simple intuitive explanations that help understanding the "physical meaning" of advanced mathematical objects.

In addition, we describe the applications of the formalism and the connections with other branches of physics. For example, we will explore the deep connection between the Brownian motion, originally introduced to describe the motion of pollen grains inside water, and quantum mechanics in Feynman's path integral formulation.

Furthermore, we describe selected important applications of the formalism of stochastic processes to various branches of modern physics, namely partial differential equations, quantum mechanics of interacting particles, and econophysics. The proposed applications are remarkable examples of complex systems, as they model a large number of strongly interacting constituents. Their study requires and applies the same mathematical notions and instruments, ranging from the Itô integral to the Feynman–Kac and Fokker–Planck equations, on which this flourishing branch of applied mathematics is based.

We also provide fully solved exercises meant to allow the reader to deepen his knowledge of the several topics; sometimes, in the exercises we take the opportunity to present material that is not covered in the main text.

While reading this book, a reader with basic knowledge of a programming language like C++ or Fortran will become able to implement his/her own simulation algorithms in computational physics, thus having the possibility to address the study of models in statistical physics and quantum mechanics. Moreover, and maybe even more importantly, the reader will develop the background necessary to consciously and critically use already existing computational physics codes.

The book is organized as follows: in Chap. 1, we will review the basic formalism of probability theorem, putting emphasis on the topics that will be more useful for the following chapters. The material in this chapter is naturally standard, but we find it useful to make the book self-contained, and, moreover, to fix the basic notations and set up the basis of the formalism in a somehow familiar context. We take the opportunity to introduce in this chapter also topic that are very interesting for a physicist, like quantum statistics, Gaussian integration, and Wick's theorem.

In Chap. 2, we will provide a brief but self-contained and rigorous review of the basic notions and results of mathematical statistics: this is a large field that deserves a full book for a comprehensive review. In our context, we find essential to present what is needed to be able to analyze experimental data or numerical data on a fully rigorous basis. Particular emphasis is put on the definition and the properties of estimators, and on hypothesis tests.

Chapter 3 deals with conditional expectation: this is a crucial ingredient of the theory of stochastic processes. Since it is quite a difficult notion, we try to add explanations and elementary constructions that allow the reader to capture the essential interpretation. After the first three chapters, which somehow provide the foundation of the book, we start dealing with stochastic processes, which are the most natural models of phenomena whose time dependence is, to a certain degree, non-deterministic.

In Chap. 4, we introduce the Markov chains, which are the simplest stochastic processes. Although simple, those processes, defined by the very intuitive picture that the state at present time is enough to study the future state, allowing to "forget the past," are extremely important in several areas of applied sciences. We focus our attention on the thermalization process of the Markov chains, that is, on the infinite time limit. This has deep connection with the thermalization of real physical systems. We will also prove the Metropolis theorem, which, as mentioned above, provides the mathematical foundation of a vast number of numerical simulations in computational physics.

Chapter 5 is meant to teach the reader how to use a computer to design his own simulations relying on what he learned till now. He/She will learn how to sample a random variable and a Markov chain, thus learning the basics of Monte Carlo simulations. This will allow him/her to numerically study a huge variety of physical models like, just to mention a couple of examples, the Ising model or the model of a simple liquid in thermal equilibrium.

Chapter 6 deals with the celebrated Brownian motion, starting from the historical approach invented by Einstein, up to the modern formulation in the context of the theory of stochastic processes. We stress that the Brownian motion is a meeting point of several branches of physics, ranging from quantum mechanics in the path integral formulation to the theory of partial differential equations.

The two final chapters will be devoted to more advanced topics, that is, stochastic calculus and stochastic differential equations, with physical applications. Attention will be devoted both to the formal development and to the applications, in the attempt of justifying the mathematical objects with their interpretation in the realm of models of physical systems. We will arrive at the Feynman–Kac and Fokker–Planck equations, which play a crucial role in several branches of physics.

Chapter 1
Review of Probability Theory

Abstract This chapter provides a self-contained review of the foundations of probability theory, in order to fix notations and introduce mathematical objects employed in the remaining chapters. In particular we stress the notions of measurability, related to what it is actually possible to observe when performing an experiment, and statistical independence. We present several tools to deal with random variables; in particular we focus on the normal random variables, a cornerstone in the theory of random phenomena. We conclude presenting the law of large numbers and the central limit theorem.

Keywords Probability theory · Probability spaces · Random variables
Probability laws · Law of large numbers · Central limit theorem

1.1 Probability Spaces and Random Variables

Let us begin by introducing the natural environment to deal with random phenomena, relying on the axiomatic formulation due to Kolmogorov [1]. The first ingredient we need is provided by the following

Definition 1.1 A **probability space** is a triplet of the form:

$$(\Omega, \mathcal{F}, P) \tag{1.1}$$

where:

1. Ω is a non–empty set, called **sample space**;

2. $\mathcal{F}$ is a σ-field of subsets of Ω, called the **set of events**;
3. P is a measure $\mathcal{F}$ satisfying the condition $P(\Omega) = 1$, called **probability measure**.

© Springer International Publishing AG, part of Springer Nature 2018
E. Vitali et al., *Theory and Simulation of Random Phenomena*, UNITEXT
for Physics, https://doi.org/10.1007/978-3-319-90515-0_1

The elements of Ω are interpreted as all the possible outcomes of an experiment modelled by the probability space $(\Omega, \mathcal{F}, P)$. Once observed the outcome of the given experiment, we know whether some events have happened or not: the collection of such events is $\mathcal{F}$. The probability measure describes how likely is the outcoming of the events. The mathematical requirement of $\mathcal{F}$ being a σ-field means, by definition, that $\mathcal{F}$ is closed under countable unions and intersections, and under complementations; moreover, Ω itself and the empty set $\emptyset$ belongs to $\mathcal{F}$ by definition. The measure P is σ-additive, i.e., for any countable family of events $\{A_1, \dots, A_n, \dots\} \subset \mathcal{F}$ such that $A_i \cap A_j = \emptyset$ if $i \neq j$:

$$P \left(\bigcup_{j=1}^{+\infty} A_j \right) = \sum_{j=1}^{+\infty} P(A_j) \tag{1.2}$$

In general, the sample space can be any non-empty set, finite, infinite countable or uncountable. Whenever Ω is finite or countably infinite, the σ-field is always taken to be the whole power set of Ω, $\mathcal{F} = \mathcal{P}(\Omega)$, containing all the subsets of Ω. Whenever Ω is uncountable, the power set is in general too large, giving rise to pathological situations in which a probability measure cannot be defined (the reader may remember that this is the case in Lebesgue measure theory). In such cases one has to restrict the σ-field. Whenever Ω is a topological space, we will always choose the **Borel σ-field**, $\mathcal{B}(\Omega)$, which is the smallest σ-field containing all the open subsets of Ω.

The second basic ingredient is the definition of random variable.

Definition 1.2 Let $(E, \mathcal{E})$ be a measurable space, that is a non-empty set E together with a σ-field of subsets $\mathcal{E}$. A **random variable** is a function $X : \Omega \to E$ which is **measurable**, that is:

$$\forall B \in \mathcal{E}, \quad \{\omega \in \Omega : X(\omega) \in B\} \in \mathcal{F} \tag{1.3}$$

When we want to make explicit the σ-fields, we will use the transparent notation:

$$X : (\Omega, \mathcal{F}, P) \to (E, \mathcal{E}) \tag{1.4}$$

Remark 1.1 We will use simple notation of the form $\{X \in B\}$ instead of the more precise one $\{\omega \in \Omega : X(\omega) \in B\}$, or, in case of real valued random variables, $\{X \leq x\}$ instead of $\{\omega \in \Omega : X(\omega) \leq x\}$ and so on.

Random variables are thus the functions naturally related to the mathematical structure of measurable spaces, just like continuous functions between topological spaces and linear applications between vector spaces.

The map:

$$B \in \mathcal{E} \to \mu(B) \stackrel{def}{=} P(X \in B) \tag{1.5}$$

is a probability measure on $\mathcal{E}$ and is called the **law** or **distribution** of the random variable X. The notation $X \sim \mu$ is commonly use to denote the fact that μ is the law of X. The measurable space $(E, \mathcal{E})$, equipped with the measure μ, becomes a probability space $(E, \mathcal{E}, \mu)$.

Remark 1.2 We note that the identity map on $(\Omega, \mathcal{F})$:

$$\omega \in \Omega \rightarrow id_\Omega(\omega) = \omega \tag{1.6}$$

is naturally a random variable and its law is precisely P. This simple observation allows to conclude that, whenever a probability measure P is defined on a measurable set, there always exist a random variable taking values in the given set whose law is P. This could appear trivial at first sight, but it is useful: we will always work directly with laws, forgetting to explicitly define the random variables.

We stress that the notion of **measurability** is very important: from the point of view of the interpretation, the fact that a random variable X is measurable means that, once observed the outcome of an experiment modelled by $(\Omega, \mathcal{F}, P)$, the value of X is known. This will turn out to be a key point in future chapters, when our knowledge will depend on time.

Besides measurability, an extremely important notion is that of **independence**, translating in mathematical language the intuitive idea the term suggests.

Definition 1.3 A collection $\{\mathcal{F}_j\}_{j \in \mathcal{J}}$ (not necessarily finite) of sub-σ-fields of $\mathcal{F}$ are said to be **independent** if, for any finite subset $\mathcal{I} \subset \mathcal{J}$ the following equality holds:

$$P\left(\bigcap_{i \in \mathcal{I}} A_i\right) = \prod_{i \in \mathcal{I}} P(A_i) \tag{1.7}$$

for any choice of events $A_i \in \mathcal{F}_i$.

A collection $\{A_j\}_{j \in \mathcal{J}}$ of events belonging to $\mathcal{F}$ are said to be **independent** if the sub-σ-fields $\{\mathcal{F}_{A_j}\}_{j \in \mathcal{J}}$, $\mathcal{F}_{A_j} = \{\emptyset, A_j, A_j^C, \Omega\}$, are independent.

A collection $\{X_j\}_{j \in \mathcal{J}}$ of random variables, $X_j : (\Omega, \mathcal{F}, P) \rightarrow (E_j, \mathcal{E}_j)$, are said to be **independent** if the **generated** sub-σ-fields $\{\sigma(X_j)\}_{j \in \mathcal{J}}$, $\sigma(X_j)$ being, by definition, the smallest σ-field containing all the events $\{X_j \in B\}$ for all $B \in \mathcal{E}_j$, are independent.

Remark 1.3 The definition of a σ-field $\sigma(X)$ generated by a random variable X, will be useful in the following chapters. It is the smallest σ-field making X a measurable function. This means that it contains the minimal information needed to know the value of X, once an experiment is performed.

1.2   First Examples: Binomial Law, Poisson Law, and Geometric Law

Now that we have introduced the most important ingredients, we are ready to build up our first examples of probability spaces and random variables. We consider an experiment consisting in tossing a (non necessarily balanced) coin n times and counting the number of heads obtained. How can we describe such situation in the language of probability theory? It is quite natural to build up a probability space $(\Omega, \mathcal{F}, P)$ in the following way; let's choose:

$$\Omega = \{0, 1\}^n \tag{1.8}$$

This means that the possible outcomes have the form $\Omega \ni \omega = (\omega_1, \ldots, \omega_n)$ where, using a simple convention, $\omega_i = 0$ if at the i-th toss we get tail and $\omega_i = 1$ if we get head. We may consider as σ-field the whole power set of Ω, $\mathcal{F} = \mathcal{P}(\Omega)$, containing all the subsets of Ω.

The definition of the probability measure requires some more work. We introduce a parameter $p \in [0, 1]$, which would be equal to $1/2$ if the coin were perfectly balanced, with the interpretation of how likely is the outcome of head in a single toss. Rigorously, this means we are defining:

$$P(A_i) = p, \quad \forall i = 1, \ldots, n, \quad A_i = \{\omega = (\omega_1, \ldots, \omega_n) \in \Omega \mid \omega_i = 1\} \tag{1.9}$$

assuming that all tosses are *equivalent*, i.e. $P(A_i)$ does not depend on i. We observe now that the event "the first x tosses ($x = 0, \ldots, n$) give head and the other $(n - x)$ tail", contains only the element of Ω:

$$\omega = (1, 1, \ldots, 1, 0, \ldots, 0) = A_1 \cap \cdots \cap A_x \cap A_{x+1}^C \cap \cdots \cap A_n^C \tag{1.10}$$

If we assume that the tosses are **independent** we have necessarily:

$$P(\{\omega = (1, 1, \ldots, 1, 0, \ldots, 0)\}) = p^x (1 - p)^{n-x} \tag{1.11}$$

Moreover any element of ω in which the value 1 appears x times and the value 0 $(n - x)$ times has the same probability by construction. We have thus defined $P(\{\omega\})$ for all $\omega \in \Omega$ and thus:

$$P(A) = \sum_{\omega \in A} P(\{\omega\}) \tag{1.12}$$

for any event $A \in \mathcal{F} = \mathcal{P}(\Omega)$.

The definition of the probability space $(\Omega, \mathcal{F}, P)$ is now completed.

Let's define now the random variable $X : \Omega \rightarrow \{0, 1, \ldots, n\}$:

$$\omega = (\omega_1, \ldots, \omega_n) \rightarrow X(\omega) = \sum_{i=1}^{n} \omega_i \qquad (1.13)$$

where the set $\{0, 1, \ldots, n\}$ is trivially measurable once endowed with its power set σ-field $\mathcal{P}(\{0, 1, \ldots, n\})$. The random variable X simply counts the number of heads in n independent coin tosses. The law of X can be obtained very simply starting from the probabilities:

$$P(X = x), \quad x \in \{0, 1, \ldots, n\} \qquad (1.14)$$

that can be obtained counting the number of different $\omega \in \Omega$ in which the value 1 appears x times and the value 0 $(n - x)$ times: the event

$$\{X = x\} = \{\omega \in \Omega \mid \sum_{i=1}^{n} \omega_i = x\} \qquad (1.15)$$

contains indeed all such elements, whose number is given by the binomial coefficient:

$$\binom{n}{x} = \frac{n!}{x!(n - x)!}, \quad n! = n(n - 1)(n - 2) \ldots 1 \qquad (1.16)$$

The result is:

$$P(X = x) = \binom{n}{x} p^x (1 - p)^{n-x}, \quad x = 0, \ldots, n \qquad (1.17)$$

From the knowledge of $P(X = x)$ we immediately obtain the law of X:

$$B \in \mathcal{P}(\{0, 1, \ldots, n\}) \rightarrow \mu(B) = \sum_{x \in B} \binom{n}{x} p^x (1 - p)^{n-x} \qquad (1.18)$$

This law is very famous and it is called **binomial law with parameters** (n, p): we will write $X \sim B(n, p)$. In the particular case $n = 1$, $B(1, p)$ is called **Bernoulli law with parameter** p.

In order to make the notations more compact, it is useful to define a function $p : \mathbb{R} \rightarrow \mathbb{R}$ as follows:

$$x \in \mathbb{R} \rightarrow p(x) = \begin{cases} \binom{n}{x} p^x (1 - p)^{n-x} & \text{if } x = 0, \ldots, n \\ 0 & \text{otherwise} \end{cases} \qquad (1.19)$$

Such function is called **discrete density** of X, it is different from zero only in a countable subset of $\mathbb{R}$ and satisfies:

$$p(x) \geq 0, \quad \sum_{x \in \mathbb{R}} p(x) = 1 \tag{1.20}$$

where the sum is meant in the language of infinite summations theory. The law can be extended naturally to the measurable set made by real numbers equipped with the Borel σ-field $\mathcal{B}(\mathbb{R})$, the smallest σ-field containing the open subset of $\mathbb{R}$, $(\mathbb{R}, \mathcal{B}(\mathbb{R}))$

$$B \in \mathcal{B}(\mathbb{R}) \to \mu(B) = \sum_{x \in B} p(x) \tag{1.21}$$

In this example we have built up explicitly a probability space $(\Omega, \mathcal{F}, P)$ and a **discrete** random variable X (i.e. it assumes only a countable set of values). The law of such random variable turned out to be completely determined by the discrete density $p(x)$. The reader will imediately realize that the precise details of the definition of the space $(\Omega, \mathcal{F}, P)$ and of X can be completely forgotten once the law of X is known: they actually have no impact on the probabilistic description of the experiment.

Starting from the binomial law, it is possible to build up other very important laws. We consider a law $B(n, \frac{\lambda}{n})$, where $\lambda > 0$ is a fixed parameter, and we investigate the asymptotic behavior as $n \to +\infty$:

$$P(X = x) = \binom{n}{x} \left(\frac{\lambda}{n}\right)^x \left(1 - \frac{\lambda}{n}\right)^{n-x} = \frac{n!}{x!(n-x)!} \frac{\lambda^x}{n^x} \left(1 - \frac{\lambda}{n}\right)^{n-x} = \tag{1.22}$$

$$= \frac{\lambda^x}{x!} \left(1 - \frac{\lambda}{n}\right)^n \frac{n(n-1)\ldots(n-x+1)}{n^x} \left(1 - \frac{\lambda}{n}\right)^{-x} \stackrel{n \to +\infty}{\longrightarrow} \frac{\lambda^x}{x!} e^{-\lambda}$$

The last expression provides the definition of the **Poisson law with parameter** λ, related to the discrete density:

$$x \in \mathbb{R} \to p(x) = \begin{cases} \frac{\lambda^x}{x!} e^{-\lambda} & x = 0, 1, 2, \ldots \\ 0 & otherwise \end{cases} \tag{1.23}$$

It is simple to check that the above function actually defines a discrete density:

$$\sum_{x \in \mathbb{R}} p(x) = \sum_{x=0}^{+\infty} \frac{\lambda^x}{x!} e^{-\lambda} = e^{\lambda} e^{-\lambda} = 1 \tag{1.24}$$

A Poisson law is often used to model experiments in which a system of many objects is observed (for example a collection of nuclei), each having a very low probability to undergo a certain phenomenon (for example radioactive decay).

Another interesting question is the following: what is the probability that the first head appears precisely at the x-th toss? Let T be the random variable providing the toss in which we obtain the first head. To evaluate $P(T = x)$ we can use the following simple identity:

$$\{T = x\} \cup \{T > x\} = \{T > x - 1\} \tag{1.25}$$

implying, since naturally $\{T = x\} \cap \{T > x\} = \emptyset$:

$$P(T = x) + P(T > x) = P(T > x - 1) \tag{1.26}$$

The key point is that the event $\{T > x\}$ corresponds to no heads in the first x tosses, so that the probability is:

$$P(T > x) = \binom{x}{0} p^0 (1 - p)^{x-0} = (1 - p)^x \tag{1.27}$$

It follows that:

$$P(T = x) = P(T > x - 1) - P(T > x) = (1 - p)^{x-1} - (1 - p)^x = p(1 - p)^{x-1} \tag{1.28}$$

We call **geometric law of parameter** $p \in [0, 1]$ the law associated with the discrete density:

$$x \in \mathbb{R} \to p(x) = \begin{cases} p(1 - p)^x & x = 0, 1, 2, \ldots \\ 0 & otherwise \end{cases} \tag{1.29}$$

In our example, $T - 1$ has a geometric law.

1.3 Probability and Counting

Within probability theory, quite often it is necessary to be able to count the elements of a finite set. Whenever the sample space Ω is finite, $\Omega = \{\omega_1, \ldots, \omega_N\}$, one uniquely defines the probability measure P assigning the numbers:

$$p_i = P(\{\omega_i\}), \quad i = 1, \ldots, N \tag{1.30}$$

For any event $A \in \mathcal{P}(\Omega)$, its probability is given by:

$$P(A) = \sum_{\omega_i \in A} p_i \tag{1.31}$$

In the particular case of **uniform probability** (we use the symbol $card(X)$ to denote the number of elements of a set X):

$$p_i = \frac{1}{card(\Omega)} = \frac{1}{N} \tag{1.32}$$

we have:

$$P(A) = \frac{card(A)}{card(\Omega)} \tag{1.33}$$

This relation should be familiar to all readers and reminds the intuitive idea of probability as the ratio between the number of favorable events and the total number of possible events.

Moreover, the above relation indicates that, when facing probability problems, a good skill in counting is desirable. The field of mathematics that deals with counting the number of elements of a set is combinatorics. We won't enter the details of such highly non trivial branch of mathematics. We just summarize a few commonly used notations. The factorial of a non-negative integer n is defined as:

$$n! = n\,(n-1)\ldots 1 = \prod_{i=1}^{n} i, \quad 0! = 1 \tag{1.34}$$

and is the number of bijective functions $f : \{1, \ldots, n\} \to \{1, \ldots, n\}$, that is the number of permutations of n objects. Another very important object is the binomial coefficient, we have already encountered in the previous section:

$$\binom{n}{k} = \frac{n!}{k!(n-k)!} \tag{1.35}$$

which can be defined as the coefficient of x^k in the expansion of $(1 + x)^n$. $\binom{n}{k}$ is the number of subsets containing k elements that can be extracted from the set $\{1, \ldots, n\}$.

1.3.1 A Bit of Quantum Statistics

We find interesting to present a counting exercise that deserves strong attention from an historical point of view, in connection with the origins of quantum mechanics. Let's introduce the problem step by step.

Suppose a single particle energy level ε has degeneracy g (this means that there are g energy levels sharing the same energy ε) and, in an ideal quantum gas, n particles can occupy any of the degenerate states. We wish to count the number of different ways we can distribute the n particles within the g degenerate states.

Actually the solution depends on a crucial point in quantum mechanics and statistical physics: distinguishability and indistinguishability.

If the particles are assumed to be **distinguishable**, naturally, starting with the first particle we have g possibilities to place it inside a single particle state, and so on with the other particles. The desired number is thus (**Maxwell-Boltzmann counting**):

$$\mathcal{N} = g^n \tag{1.36}$$

On the other hand, if the particles are **indistinguishable**, the counting is strongly different. When the particles are **bosons**, there are no restrictions on the number of particles that can occupy a single particle state, and thus we can proceed as follows: let's consider n particles and $g-1$ separating walls among the levels (the reader may think of the levels as boxes and of the particles as balls to be distributed among the boxes). We have thus $n+g-1$ objects, and the number of ways we are looking for is simply the number of ways we can choose n of these objects to be the particles, that is (**Bose-Einstein counting**):

$$\aleph = \binom{n+g-1}{n} = \frac{(n+g-1)!}{(g-1)!n!} \qquad (1.37)$$

Finally, if the particles are **fermions**, Pauli principle imposes a severe restriction: in each single-particle state there can be at most one particle. Assuming $g > n$, we can perform the counting as follows. We have to decide which single particle states contain one particle and which are empty. The number of choices is (**Fermi-Dirac counting**):

$$\aleph = \binom{g}{n} = \frac{g!}{(g-n)!n!} \qquad (1.38)$$

Suppose now we have a collection, possibly infinite, of single particle energy levels $\{\varepsilon_\alpha\}_\alpha$, with degeneracies g_α. We would like to evaluate the number of ways we can distribute N particles inside the levels in such a way that n_α particles have energy ε_α. Naturally $N = \sum_\alpha n_\alpha$.

Let's use the notation $t(\{n_\alpha\})$ for the desired number of **microstates** corresponding to the desired partition of the particles among the energy levels. In what follows, we will deal with bosons and fermions, forgetting about distinguishable particles. Relying on the previous results, for bosons we have:

$$t(\{n_\alpha\}) = \prod_\alpha \frac{(n_\alpha + g_\alpha - 1)!}{(g_\alpha - 1)!n_\alpha!} \qquad (1.39)$$

while, for fermions, we have:

$$t(\{n_\alpha\}) = \prod_\alpha \frac{g_\alpha!}{(g_\alpha - n_\alpha)!n_\alpha!} \qquad (1.40)$$

As usual in statistical physics, we claim that, in thermal equilibrium, the particles populate the levels, on average, in such a way that, given an average energy $E = \sum_\alpha n_\alpha \varepsilon_\alpha$ and an average particles number $N = \sum_\alpha n_\alpha$, the number of microstates corresponding to the given average distribution is maximum.

Let's perform in detail the calculations in the bosonic case, the fermionic one being very similar. We will assume high degenaracy and let:

$$t\left(\{n_\alpha\}\right) \simeq \prod_\alpha \frac{(n_\alpha + g_\alpha)!}{g_\alpha! n_\alpha!} \tag{1.41}$$

Moreover, we (brutally!) assume that (a simplified form of the) Stirling formula can be applied and write:

$$\log\left(t\left(\{n_\alpha\}\right)\right) \simeq \sum_\alpha (n_\alpha + g_\alpha)\log(n_\alpha + g_\alpha) - g_\alpha \log(g_\alpha) - n_\alpha \log(n_\alpha) \tag{1.42}$$

To find the optimal partition, we find the stationary point of the function:

$$\log\left(t\left(\{n_\alpha\}\right)\right) + a\sum_\alpha n_\alpha + b\sum_\alpha n_\alpha \varepsilon_\alpha \tag{1.43}$$

where a, b are Lagrange multipliers, fixing the average particle number and the average total energy. We thus find the equation:

$$\log(n_\alpha + g_\alpha) + 1 - \log(n_\alpha) - 1 + a + b\varepsilon_\alpha = 0 \tag{1.44}$$

that is:

$$\frac{n_\alpha + g_\alpha}{n_\alpha} = \exp\left(-a - b\varepsilon_\alpha\right) \tag{1.45}$$

or:

$$n_\alpha = \frac{g_\alpha}{\exp\left(-a - b\varepsilon_\alpha\right) - 1} \tag{1.46}$$

where a, b are determined by the constraints:

$$N = \sum_\alpha n_\alpha, \quad E = \sum_\alpha n_\alpha \varepsilon_\alpha \tag{1.47}$$

Typically the optimal partition is written in terms of **temperature** $\beta = \frac{1}{k_B T}$ and **chemical potential** μ in the following way: (**Bose-Einstein distribution**):

$$n_\alpha = \frac{g_\alpha}{\exp\left(\beta(\varepsilon_\alpha - \mu)\right) - 1} \tag{1.48}$$

In the fermionic case, the same procedure leads to the following optimal partition (**Fermi-Dirac distribution**):

$$n_\alpha = \frac{g_\alpha}{\exp\left(\beta(\varepsilon_\alpha - \mu)\right) + 1} \tag{1.49}$$

These distributions are very famous in statistical physics. For example, Bose-Einstein distribution is the key to understand the black-body radiation and Bose-Einstein condensation, while the Fermi-Dirac distribution puts light into the description of the behavior of electrons in metals.

1.4 Absolutely Continuous Random Variables

In the preceding examples we have presented our first examples of random variables. All such examples involved real valued **discrete** random variables, taking values in a countable subset of $\mathbb{R}$; their law is univocally determined by the discrete density $p(x)$, non-zero only inside a countable set, non-negative and normalized to one. The law has the form:

$$B \in \mathcal{B}(\mathbb{R}) \rightarrow \mu(B) = \sum_{x \in B} p(x) \tag{1.50}$$

The generalization to multidimensional **discrete** random variables is straightforward: one simply defines discrete densities $p(\mathbf{x})$ on $\mathbb{R}^d$, related to the laws of random variables of the form $X = (X_1, \ldots, X_d)$ where, naturally, the X_i are real valued discrete random variables.

In general, we will very often meet random variables which are not discrete. The simplest example is provided by the **uniform law** in $(0, 1)$: we will say that a random variable is uniform in $(0, 1)$ if its law has the form:

$$\mu(B) = \int_B dx \, p(x) \tag{1.51}$$

where:

$$p(x) = \begin{cases} 1 & x \in (0, 1) \\ 0 & x \notin (0, 1) \end{cases} \tag{1.52}$$

The values of X cover uniformly the interval $(0, 1)$. Random variables uniform in $(0, 1)$ will be very important in the definition of sampling techniques in the following chapters: the key point is that, with a computer, we can *generate* the values of X, in a sense which will be later clarified.

It is evident that, in (1.52), the discrete density has been replaced by a continuous density and the summation has been replaced by an integral. The random variable X belongs to a very important class of random variables, defined in the following definition.

Definition 1.4 We say that a random variables taking values in $\mathbb{R}^d$, $X : (\Omega, \mathcal{F}, P) \rightarrow (\mathbb{R}^d, \mathcal{B}(\mathbb{R}^d))$ is **absolutely continuous** if the law μ of X is absolutely continuous with respect to the Lebesgue measure, that is $\mu(B) = 0$ whenever B has Lebesgue measure equal to zero.

If X is absolutely continuous, the Radon–Nicodym theorem, from measure theory, ensures the existence of the **density of** X, i.e. a function $p : \mathbb{R}^d \to \mathbb{R}$ non-negative, Borel-measurable, Lebesgue-integrable with $\int_{\mathbb{R}^d} dx\, p(\mathbf{x}) = 1$, and such that:

$$\mu(B) = \int_B dx\, p(\mathbf{x}) = \int_{\mathbb{R}^d} dx\, 1_B(\mathbf{x})\, p(\mathbf{x}) \quad \forall B \in \mathcal{B}(\mathbb{R}^d) \qquad (1.53)$$

The density of a random variable is unique almost everywhere with respect to Lebesgue measure: if p e p' are two densities of a random variable X, then necessarily they coincide everywhere but inside a set of zero Lebesgue measure.

Remark 1.4 We invite the reader to observe the similarity between the discrete and the absolutely continuous case. If a random variable has discrete density $p_d(\mathbf{x})$, then:

$$\mu(B) = \sum_{\mathbf{x} \in B} p_d(\mathbf{x}) \qquad (1.54)$$

while, if it is absolutely continuous, there exist a density $p_c(\mathbf{x})$ such that:

$$\mu(B) = \int_B dx\, p_c(\mathbf{x}) \qquad (1.55)$$

Several authors unify the two cases using the integral notation defining $p_d(\mathbf{x})$ as a sum of Dirac's deltas. We prefer not to use such a notation. We observe, on the other hand, that there exist random variables which are neither discrete nor absolutely continuous.

In the case of real valued random variables, it is always possible to define the **cumulative distribution function**, $F : \mathbb{R} \to \mathbb{R}$ as follows:

$$x \in \mathbb{R} \to F(x) \stackrel{def}{=} \mu((-\infty, x]) = P(X \le x) \qquad (1.56)$$

By construction, F is increasing, right-continuous, and satisfies:

$$\lim_{x \to -\infty} F(x) = 0, \quad \lim_{x \to +\infty} F(x) = 1 \qquad (1.57)$$

It is possible to show that there is a one to one correspondence between the cumulative distribution function and the law of a random variable. Moreover, for any function F possessing the above mentioned properties, there exists a random variable having F as the cumulative distribution function.

In particular useful relations are:

$$\mu((x, y]) = F(y) - F(x), \quad \mu((x, y)) = F(y^-) - F(x^-), \ldots \qquad (1.58)$$

and:

$$\mu(\{x\}) = F(x) - F(x^-) \tag{1.59}$$

where we use the notation $F(x^-) = \lim_{t \to x^-} F(t)$.

If X is absolutely continuous we have:

$$F(x) = \int_{-\infty}^{x} dy \, p(y) \tag{1.60}$$

Moreover, if the density is continuous on $\mathbb{R}$, the cumulative distribution function is differentiable on $\mathbb{R}$ and we have:

$$p(x) = \frac{dF(x)}{dx} \tag{1.61}$$

In general (1.61) is not true for any absolutely continuous random variable, since the density can happen not to be continuous; however, it can be shown that the cumulative distribution function is always differentiable almost everywhere with respect to the Lebesgue measure, and it is always possible to modify the density in such a way that it coincides with the derivative of F in all points where such derivative exists.

Example 1.1 Our first examples of absolutely continuous random variables are the following:

1. If the density is:

$$p(x) = \begin{cases} 1, & x \in (0, 1) \\ 0, & x \notin (0, 1) \end{cases} \tag{1.62}$$

we will say that X is **Uniform** in $(0, 1)$. The cumulative distribution function of X is:

$$F(x) = \begin{cases} 0, & x \leq 0 \\ x, & x \in (0, 1) \\ 1, & x \geq 1 \end{cases} \tag{1.63}$$

2. If the density is:

$$p(x) = \frac{1}{\sqrt{2\pi}} \exp\left(-\frac{x^2}{2}\right) \tag{1.64}$$

we will say that X is **Standard Normal** and we will write $X \sim N(0, 1)$. Its cumulative distribution function is:

$$F(x) = \frac{1}{\sqrt{2\pi}} \int_{-\infty}^{x} dy \, \exp\left(-\frac{x^2}{2}\right) \tag{1.65}$$

This integral cannot be worked out analytically, but many tables and softwares are available to perform the calculation. Typically the cumulative distribution function of a standard normal random variable is expressed in terms of the famous **error function**:

$$\text{erf}(x) = \frac{2}{\sqrt{\pi}} \int_0^x du \, \exp\left(-u^2\right) \tag{1.66}$$

as:

$$F(x) = \frac{1}{2}\left(1 + \text{erf}\left(\frac{x}{\sqrt{2}}\right)\right) \tag{1.67}$$

From this relation, it simply follows that:

$$P\left(-x < X \le x\right) = P\left(|X| < x\right) = \text{erf}\left(\frac{x}{\sqrt{2}}\right) \tag{1.68}$$

Important values are $P\left(|X| < 1\right) \simeq 0.68$ and $P\left(|X| < 2\right) \simeq 0.95$.

Example 1.2 In quantum mechanics, the ground state wave function of a one-dimensional harmonic oscillator of mass $m = 1$ and elastic constant $k = 1$ is:

$$\psi(x) = \left(\frac{1}{\pi}\right)^{1/4} \exp(-\frac{1}{2}x^2) \tag{1.69}$$

According to Born interpretation of quantum mechanics, the square modulus of the wave function is the probability density for the position of the oscillator. The position of the oscillator is thus interpreted as an absolutely continuous random variable X with probability density:

$$p(x) = |\psi(x)|^2 = \left(\frac{1}{\pi}\right)^{1/2} \exp(-x^2) \tag{1.70}$$

This has again the form of a normal. We will learn to write $X \sim N(0, \frac{1}{2})$.

1.5 Integration of Random Variables

We are going now to introduce key notions to work with random variables: the expectation, the variance, and so on. We prefer to use the unifying formalism of abstract integration with respect to probability measures to introduce such notions, in order to avoid the necessity of dealing separately with discrete and continuous random variables. If a reader is not interested in abstract integration, he/she can skip directly to the formulae expressing abstract integrals in terms of ordinary integrals or summations involving the probability density.

As usual in abstract integration theories, one works with extended functions, that is random variables $X : \Omega \to \overline{R} = \mathbb{R} \cup \{\pm\infty\}$, endowing $\overline{R}$ with the Borel σ-field.

Such extension is completely innocuous, and simply meant to work with limits, superiors or inferiors extrema. The construction is very simple, and will be sketched here starting from the class of random variables introduced in the following definition.

Definition 1.5 We say that a random variable X is **simple** if it can be written as:

$$X(\omega) = \sum_{i=1}^{n} a_i \, 1_{A_i}(\omega) \tag{1.71}$$

where n is an integer number, $a_i \in \mathbb{R}$, $A_i \in \mathcal{F}$, $i = 1, \ldots, n$.

The integral of simple random variables is defined as follows:

Definition 1.6 If X is simple we define **expectation** or **abstract integral** of X with respect to the probability measure P, denoted $\int_{\Omega} X(\omega) P(d\omega)$:

$$\int_{\Omega} X(\omega) P(d\omega) \stackrel{def}{=} \sum_{i=1}^{n} a_i P(A_i) \tag{1.72}$$

If X is a non-negative random variable, we define **expectation** or **abstract integral** of X with respect to the probability measure P, denoted $\int_{\Omega} X(\omega) P(d\omega)$ the extended real number:

$$\int_{\Omega} X(\omega) P(d\omega) \stackrel{def}{=} \sup \left\{ \int_{\Omega} Y(\omega) P(d\omega) : Y \; simple, \; 0 \le Y \le X \right\} \tag{1.73}$$

which can be equal to $+\infty$.

In the most general case, we let $X^+ = \max(X, 0)$ e $X^- = -\min(X, 0)$ and introduce the following definition:

Definition 1.7 We say that $X : \Omega \to \overline{R}$ is **integrable** if $\int_{\Omega} X^+(\omega) P(d\omega) < +\infty$ and $\int_{\Omega} X^-(\omega) P(d\omega) < +\infty$. In such case we define **expectation** or **abstract integral** of X with respect to the probability measure P, and denote $\int_{\Omega} X(\omega) P(d\omega)$, the real number:

$$\int_{\Omega} X(\omega) P(d\omega) \stackrel{def}{=} \int_{\Omega} X^+(\omega) P(d\omega) - \int_{\Omega} X^-(\omega) P(d\omega) \tag{1.74}$$

We stress the important identity:

$$\int_{\Omega} 1_A(\omega) P(d\omega) = P(A), \quad \forall A \in \mathcal{F} \tag{1.75}$$

Definition 1.8 The set of **integrable** random variables $X : \Omega \to \overline{R} = \mathbb{R} \cup \{\pm\infty\}$ will be denoted $\mathcal{L}(\Omega, \mathcal{F}, P)$

Readers familiar with Lebesgue theory of integration will certainly not be surprised by the following elementary properties of the abstract integral, which will be stated without proof. The interested reader may refer to [5].

1. If X, Y are integrable random variables, $\alpha X + \beta Y$ is integrable for all $\alpha, \beta \in \mathbb{R}$ and:

$$\int_\Omega (\alpha X + \beta Y)(\omega) P(d\omega) = \alpha \int_\Omega X(\omega) P(d\omega) + \beta \int_\Omega Y(\omega) P(d\omega) \quad (1.76)$$

2. If $X \geq 0$, then $\int_\Omega X(\omega) P(d\omega) \geq 0$. If moreover $Y \geq 0$ is integrable and $0 \leq X \leq Y$, then X is integrable and $\int_\Omega X(\omega) P(d\omega) \leq \int_\Omega Y(\omega) P(d\omega)$.

3. $X \in \mathcal{L}(\Omega, \mathcal{F}, P)$ if and only if $|X| \in \mathcal{L}(\Omega, \mathcal{F}, P)$, and, in such case, we have $|\int_\Omega X(\omega) P(d\omega)| \leq \int_\Omega |X(\omega)| P(d\omega)$

4. If $X = Y$ **almost surely** (a.s.), i.e. if there exists an event N, $P(N) = 0$, such that $X(\omega) = Y(\omega)$, $\forall \omega \in N^c$, then $\int_\Omega X(\omega) P(d\omega) = \int_\Omega Y(\omega) P(d\omega)$

The following properties concern limits and approximations. A proof can be found in [5].

Theorem 1.1 *If X is non-negative, there exists a sequence $\{X_n\}_n$ of simple, non-negative random variables, such that $X_n(\omega) \leq X_{n+1}(\omega)$ for each ω and pointwise converging to X, that is:*

$$\lim_{n\to+\infty} X_n(\omega) = X(\omega), \quad \forall \omega \in \Omega \quad (1.77)$$

Theorem 1.2 (Monotone convergence theorem) *If a sequence $\{X_n\}_n$ of non-negative random variables, such that $X_n(\omega) \leq X_{n+1}(\omega)$, converges pointwise almost surely to a (non-negative) random variable X, that is:*

$$\lim_{n\to+\infty} X_n(\omega) = X(\omega) \quad a.s. \quad (1.78)$$

then:

$$\lim_{n\to+\infty} \int_\Omega X_n(\omega) P(d\omega) = \int_\Omega X(\omega) P(d\omega) \quad (1.79)$$

even if $\int_\Omega X(\omega) P(d\omega) = +\infty$. In particular, if $\lim_{n\to+\infty} \int_\Omega X_n(\omega) P(d\omega) < +\infty$, then $X \in \mathcal{L}(\Omega, \mathcal{F}, P)$.

Theorem 1.3 (Dominated convergence theorem) *If a sequence $\{X_n\}_n$ of random variables converges almost surely to a random variable X:*

$$\lim_{n\to+\infty} X_n(\omega) = X(\omega), \quad a.s. \quad (1.80)$$

and $|X_n| \leq Y$, *for all* n, *where* $Y \in \mathcal{L}(\Omega, \mathcal{F}, P)$, *then* $X_n \in \mathcal{L}(\Omega, \mathcal{F}, P)$, $X \in \mathcal{L}(\Omega, \mathcal{F}, P)$ *and:*

$$\lim_{n \to +\infty} \int_\Omega X_n(\omega) P(d\omega) = \int_\Omega X(\omega) P(d\omega) \tag{1.81}$$

Definition 1.9 We say that two random variables X and Y are equivalent if $X = Y$ almost surely. We denote $L^1(\Omega, \mathcal{F}, P)$ the set made of equivalence classes of integrable random variables:

$$\int_\Omega |X(\omega)| P(d\omega) < +\infty \tag{1.82}$$

We denote $L^2(\Omega, \mathcal{F}, P)$ the set made of equivalence classes of square-integrable random variables:

$$\int_\Omega |X(\omega)|^2 P(d\omega) < +\infty \tag{1.83}$$

In mathematics textbooks, the difference between a random variable and an equivalence class of random variables is often ignored, when this cannot give rise to confusion.

We mention the following result:

Theorem 1.4 $L^1(\Omega, \mathcal{F}, P)$ *and* $L^2(\Omega, \mathcal{F}, P)$ *are linear vector spaces, satisfying* $L^2(\Omega, \mathcal{F}, P) \subset L^1(\Omega, \mathcal{F}, P)$; *if* $X \in L^2(\Omega, \mathcal{F}, P)$ *then:*

$$\left(\int_\Omega X(\omega) P(d\omega) \right)^2 \leq \int_\Omega X(\omega)^2 P(d\omega) \tag{1.84}$$

Moreover, if $X, Y \in L^2(\Omega, \mathcal{F}, P)$, *their product is integrable* $XY \in L^1(\Omega, \mathcal{F}, P)$ *and the* (**Cauchy-Schwarz inequality**) *holds:*

$$\left| \int_\Omega X(\omega) Y(\omega) P(d\omega) \right| \leq \sqrt{ \int_\Omega X(\omega)^2 P(d\omega) \int_\Omega Y(\omega)^2 P(d\omega) } \tag{1.85}$$

Let's turn to a useful consequence of the properties of abstract integrals:

Theorem 1.5 (Chebyshev inequality) *If* $X \in L^2(\Omega, \mathcal{F}, P)$, *for all* $a > 0$ *the following inequality holds:*

$$P\left(|X| \geq a\right) \leq \frac{\int_\Omega X(\omega)^2 P(d\omega)}{a^2} \tag{1.86}$$

Proof From the obvious inequality $X^2(\omega) \geq a^2 1_{|X| \geq a}(\omega)$, it follows that:

$$\int_\Omega X(\omega)^2 P(d\omega) \geq \int_\Omega a^2 1_{|X| \geq a}(\omega) P(d\omega) = a^2 \, P \, (|X| \geq a) \qquad (1.87)$$

which is just the statement of the theorem.

We will often use the notation:

$$E \, [X] \overset{def}{=} \int_\Omega X(\omega) P(d\omega) \qquad (1.88)$$

when no confusion can rise about the probability space over which we are integrating.

We stress the identity:

$$E[1_A] = P(A), \quad \forall A \in \mathcal{F} \qquad (1.89)$$

which will be frequently used.

Definition 1.10 If $X \in L^2(\Omega, \mathcal{F}, P)$, we call **variance** of X and denote $Var(X)$ the non-negative real number:

$$Var(X) \overset{def}{=} E \left[(X - E[X])^2 \right] \qquad (1.90)$$

The Chebyshev inequality implies the following:

$$P \, (|X - E[X]| \geq a) \leq \frac{Var(X)}{a^2} \qquad (1.91)$$

which provides an interpretation of the variance: $Var(X)$ controls the dispersion of the values of X around the expectation $E[X]$.

1.5.1 Integration with Respect to the Law of a Random Variable

Let $X : (\Omega, \mathcal{F}, P) \to (E, \mathcal{E})$ be a random variable, and μ its law, i.e. $X \sim \mu$. We know that $(E, \mathcal{E}, \mu)$ is a probability space. Let now $h : (E, \mathcal{E}, \mu) \to (\mathbb{R}, \mathcal{B}(\mathbb{R}))$ be a measurable function. Then $h \circ X$ is a composition of measurable functions and therefore a real random variable. We are going to prove now the following very important theorem:

Theorem 1.6 *If $h \geq 0$ the following equality holds:*

$$\int_\Omega (h \circ X)(\omega) P(d\omega) = \int_E h(x) \mu(dx) \qquad (1.92)$$

even when both members are equal to $+\infty$. *Moreover, for any* h, $h \circ X \in L^1(\Omega, \mathcal{F}, P)$ *if and only if* $h \in L^1(E, \mathcal{E}, \mu)$ *and, in such case, the above written equality holds.*

Proof We preliminarily observe that both the members make sense, being two abstract integrals on two different probability spaces. Now, we fix $B \in \mathcal{E}$ and we remind the reader the definition of the law of X:

$$\mu(B) = P(X \in B) \tag{1.93}$$

On the other hand, we have:

$$P(X \in B) = \int_\Omega 1_{X \in B}(\omega) P(d\omega) = \int_\Omega 1_B(X(\omega)) P(d\omega) \tag{1.94}$$

and:

$$\mu(B) = \int_E 1_B(x) \mu(dx) \tag{1.95}$$

so that the statement of the theorem is true if $h(x) = 1_B(x)$. By linearity, the equality holds also when h is a simple function.

Let now $h \geq 0$; we know that there exists a sequence $\{h_n\}_n$ of non-negative simple functions, satisfying $h_n(x) \leq h_{n+1}(x)$, and pointwise converging to h. Then, the monotone convergence theorem, applied twice, justifies the following chain of equalities:

$$\int_E h(x)\mu(dx) = \int_E \lim_{n\to\infty} h_n(x)\mu(dx) = \tag{1.96}$$
$$= \lim_{n\to\infty} \int_E h_n(x)\mu(dx) = \lim_{n\to\infty} \int_\Omega (h_n \circ X)(\omega) P(d\omega) =$$
$$= \int_\Omega \lim_{n\to\infty} h_n \circ X P(d\omega) = \int_\Omega (h \circ X)(\omega) P(d\omega)$$

proving the theorem in the case $h \geq 0$. Finally, if we consider $|h|$, we immediately conclude that $h \circ X \in L^1(\Omega, \mathcal{F}, P)$ if and only if $h \in L^1(E, \mathcal{E}, \mu)$, and the equality between the abstract integrals follows writing $h = h^+ + h^-$.

Now, we focus on the particular case $(E, \mathcal{E}) = (\mathbb{R}^d, \mathcal{B}(\mathbb{R}^d))$; moreover, we assume X absolutely continuous. If $h = 1_B$ we have:

$$P(X \in B) = \int_\Omega 1_B(X(\omega)) P(d\omega) = \mu(B) = \int_B d\mathbf{x}\, p(\mathbf{x}) = \int_{\mathbb{R}^d} d\mathbf{x}\, 1_B(\mathbf{x})\, p(\mathbf{x})$$
$$\tag{1.97}$$

where the last integrals are ordinary Lebesgue integrals. We can extend the above result to simple h exploiting linearity and to generic h using the monotone convergence theorem, obtaining the useful identity:

$$\int_\Omega (h \circ X)(\omega) P(d\omega) = \int_{\mathbb{R}^d} d\mathbf{x}\, h(\mathbf{x})\, p(\mathbf{x}) \tag{1.98}$$

which holds for any $h \geq 0$ and for any h such that the two integrals exist.

If in particular $d = 1$, $h(x) = x$ and $X \in L^1(\Omega, \mathcal{F}, P)$, we have:

$$E[X] = \int_\Omega X(\omega) P(d\omega) = \int_{-\infty}^{+\infty} dx\, x\, p(x) \qquad (1.99)$$

Moreover, if $X \in L^2(\Omega, \mathcal{F}, P)$, we have:

$$Var(X) = \int_{-\infty}^{+\infty} dx\, (x - E[X])^2\, p(x) \qquad (1.100)$$

If the random variable X is **discrete** and $p(\mathbf{x})$ is its discrete density, we have:

$$P(X \in B) = \int_\Omega 1_B(X(\omega)) P(d\omega) = \mu(B) = \sum_{\mathbf{x} \in B} p(\mathbf{x}) = \sum_{\mathbf{x} \in \mathbb{R}^d} 1_B(\mathbf{x}) p(\mathbf{x})$$

$$(1.101)$$

and thus:

$$\int_\Omega (h \circ X)(\omega) P(d\omega) = \sum_{\mathbf{x} \in \mathbb{R}^d} h(\mathbf{x}) p(\mathbf{x}) \qquad (1.102)$$

if $h \geq 0$ or if the abstract integral and the finite or infinite sum are finite. If $X \in L^1(\Omega, \mathcal{F}, P)$, we have:

$$E[X] = \int_\Omega X(\omega) P(d\omega) = \sum_{x \in \mathbb{R}} x\, p(x) \qquad (1.103)$$

and if $X \in L^2(\Omega, \mathcal{F}, P)$, we have:

$$Var(X) = \sum_{x \in \mathbb{R}} (x - E[X])^2\, p(x) \qquad (1.104)$$

1.6   Transformations Between Random Variables

Let now X be a real valued random variable defined on a probability space $(\Omega, \mathcal{F}, P)$, absolutely continuous with density $p_X(x)$. Let also $g : \mathbb{R} \to \mathbb{R}$ be a Borel-measurable function. $Y = g(X)$ is naturally a random variable. It would be useful to express its law, or its density (if it exists), in terms of the law of X. We start from the cumulative distribution function:

$$F_Y(y) = P(Y \leq y) = P(g(X) \leq y) = P(X \in B_y) = \int_{B_y} dx\, p_X(x) \qquad (1.105)$$

where:

$$B_y = \{x \in \mathbb{R} : g(x) \le y\} \qquad (1.106)$$

Let us examine first the simple case in which $g : \mathbb{R} \to \mathbb{R}$ is bijective, differentiable with continuous derivative with $\frac{dg}{dx} \ne 0$; in particular, we assume g strictly increasing. Then g is invertible on $\mathbb{R}$ with inverse g^{-1} which is differentiable with continuous non-vanishing derivative; thus:

$$B_y = \{x \in \mathbb{R} : g(x) \le y\} = \{x \in \mathbb{R} : x \le g^{-1}(y)\} \qquad (1.107)$$

which implies that:

$$F_Y(y) = P(Y \le y) = P(g(X) \le y) = P(X \le g^{-1}(y)) = F_X(g^{-1}(y)) \quad (1.108)$$

where F_X is the cumulative distribution function of X. Under the hypotheses we have fixed about the function g, if F_X is everywhere differentiable (this happens if $p_X(x)$ is continuous), also F_Y is differentiable, ensuring the existence of the density of $Y = g(X)$:

$$p_Y(y) = \frac{d F_Y(y)}{dy} = \frac{d}{dy}(F_X \circ g^{-1})(y) = p_X(g^{-1}(y))\frac{dg^{-1}(y)}{dy} \qquad (1.109)$$

If, on the other hand, g is strictly decreasing, we have:

$$F_Y(y) = P(g(X) \le y) = P(X > g^{-1}(y)) = 1 - F_X(g^{-1}(y)) \qquad (1.110)$$

and thus:

$$p_Y(y) = \frac{d F_Y(y)}{dy} = -\frac{d}{dy}(F_X \circ g^{-1})(y) = p_X(g^{-1}(y))\left(-\frac{dg^{-1}(y)}{dy}\right) \quad (1.111)$$

Combining the above results, we have proved the following:

Theorem 1.7 *If a real random variable X has continuous density $p_X(x)$ and $g : \mathbb{R} \to \mathbb{R}$ is a bijective function, differentiable with continuous derivate never equal to zero, then the random variable $Y = g(X)$ has density given by:*

$$p_Y(y) = p_X(g^{-1}(y))\left|\frac{dg^{-1}(y)}{dy}\right| \qquad (1.112)$$

As a simple application we consider affine transformations $g(x) = \sigma x + \mu$, $\sigma, \mu \in \mathbb{R}$, $\sigma \ne 0$. Since $g^{-1}(y) = \frac{y-\mu}{\sigma}$, the theorem above provides the density of $Y = \sigma X + \mu$:

$$p_Y(y) = p_X\left(\frac{y - \mu}{\sigma}\right)\frac{1}{|\sigma|} \qquad (1.113)$$

In particular, if $X \sim N(0, 1)$ is a standard normal random variable, $Y = \sigma X + \mu$, with $\mu \in \mathbb{R}$, $\sigma \in (0, +\infty)$, has density:

$$p_Y(y) = \frac{1}{\sqrt{2\pi}\sigma} \exp\left(-\frac{(y - \mu)^2}{2\sigma^2}\right) \tag{1.114}$$

We will say that $Y = \sigma X + \mu$ is a **normal with parameters** μ **and** σ^2 and we will write $Y \sim N(\mu, \sigma^2)$.

When the hypotheses of the above proved theorem do not hold, we have to work directly with the equality:

$$F_Y(y) = P(Y \leq y) = P(g(X) \leq y) = P(X \in B_y) = \int_{B_y} dx \, p_X(x) \tag{1.115}$$

where:

$$B_y = \{x \in \mathbb{R} : \ g(x) \leq y\} \tag{1.116}$$

Example 1.3 Let $X \sim N(0, 1)$ be a standard normal and $g(x) = x^2$; we wish to evaluate the density of $Y = g(X) = X^2$. Naturally $F_Y(y)$ is zero if $y \leq 0$. On the other hand, if $y > 0$, we have:

$$F_Y(y) = P(Y \leq y) = P(X^2 \leq y) = P(-\sqrt{y} \leq X \leq \sqrt{y}) = \tag{1.117}$$
$$= F_X(\sqrt{y}) - F_X(-\sqrt{y})$$

with:

$$F_X(\pm\sqrt{y}) = \int_{-\infty}^{\pm\sqrt{y}} dx \, \frac{1}{\sqrt{2\pi}} \exp\left(-\frac{x^2}{2}\right) \tag{1.118}$$

We see that $F_Y(y)$ is almost everywhere differentiable (except at the origin), and thus we can obtain the density by differentiation, obtaining:

$$p_Y(y) = \frac{1}{\sqrt{2\pi}} \frac{\exp(-y/2)}{\sqrt{y}} 1_{(0,+\infty)}(y) \tag{1.119}$$

We say that $Y = X^2$ is **chi-square with one degree of freedom**, and we write $Y \sim \chi^2(1)$.

1.7 Multi-dimensional Random Variables

Now we consider random variables taking values in $\mathbb{R}^d$. For simplicity of exposition, we work with $d = 2$, but the results can be readily generalized to higher dimensions, with less transparent notations.

Let therefore $X = (Y, Z)$ be a two-dimensional random variable, absolutely continuous with density $p(y, z)$. An interesting result is the following, which we just state without proof:

Theorem 1.8 *Y and Z are real random variables absolutely continuous, with densities $p_Y(y)$ e $p_Z(z)$ given by:*

$$p_Y(y) = \int_{-\infty}^{+\infty} dz \, p(y, z), \quad p_Z(z) = \int_{-\infty}^{+\infty} dy \, p(y, z) \tag{1.120}$$

Moreover, Y and Z are independent if and only if:

$$p(y, z) = p_Y(y)p_Z(z) \tag{1.121}$$

almost everywhere with respect to Lebesgue measure.

The densities $p_Y(y)$ and $p_Z(z)$ are called **marginal densities** of p. We observe that, once p is known, the marginal densities can be obtained, but the converse is not true: if we know $p_Y(y)$ and $p_Z(z)$, we can obtain p only if Y and Z are independent.

If $Y, Z \in L^2(\Omega, \mathcal{F}, P)$, then the product YZ is integrable: $YZ \in L^1(\Omega, \mathcal{F}, P)$. We call **covariance** of Y and Z, and we denote $Cov(Y, Z)$, the real number:

$$Cov(Y, Z) = E\left[(Y - E[Y])(Z - E[Z])\right] = E[YZ] - E[Y]E[Z] \tag{1.122}$$

From the theorem of integration with respect to the law of a random variable, in the special case $X = (Y, Z)$, $(E, \mathcal{E}) = (\mathbb{R}^2, \mathcal{B}(\mathbb{R}^2))$ and $h(y, z) = yz$, we obtain the following identity:

$$E[YZ] = \int_{\mathbb{R}^2} dydz \, yz \, p(y, z) \tag{1.123}$$

If Y and Z are independent, then:

$$E[YZ] = \int_{\mathbb{R}^2} dydz \, yz \, p_Y(y)p_Z(z) = E[Y]E[Z] \tag{1.124}$$

thanks to Fubini theorem from Lebesgue integral theory, and thus:

$$Cov(Y, Z) = 0 \tag{1.125}$$

The property of having null covariance is called **non-correlation**. We have proved that two independent random variables are non-correlated; the converse, in general is not true. A simple example of two non-correlated, non independent random variables is provided by a standard normal random variable $X \sim N(0, 1)$ and its square X^2. Obviously X, X^2 are not independent, but

$$E[XX^2] - E[X]E[X^2] = 0 \tag{1.126}$$

and thus they are also non-correlated.

We call **correlation coefficient** of Y and Z the real number:

$$\rho_{YZ} = \frac{Cov(YZ)}{\sqrt{Var(Y)}\sqrt{Var(Z)}} \tag{1.127}$$

This is zero if Y and Z are non-correlated, and, in general, satifies the following property:

$$-1 \le \rho_{YZ} \le 1 \tag{1.128}$$

which is a simple consequence of Cauchy-Schwarz inequality.

1.7.1   Evaluation of Laws

Let $X = (Y, Z)$, as before, a two–dimensional random variable absolutely continuous, with density $p(y, z)$. We wish to evaluate the law of $Y + Z$, a real random variable.

Let's evaluate the cumulative distribution function:

$$F(u) = P(Y + Z \le u) = P(X \in A_u) = \int_{A_u} dy dz \, p(y, z) \tag{1.129}$$

where:

$$A_u = \{(y, z) \in \mathbb{R}^2 : y + z \le u\} \tag{1.130}$$

Then:

$$F(u) = \int_{-\infty}^{+\infty} dy \int_{-\infty}^{u-y} dz \, p(y, z) \tag{1.131}$$

With a change of variables $z \to z' = z + y$ we have:

$$F(u) = \int_{-\infty}^{u} dz' \int_{-\infty}^{+\infty} dy \, p(y, z' - y) \tag{1.132}$$

which provides the expression for the density of the sum of two random variables:

$$p_{Y+Z}(u) = \int_{-\infty}^{+\infty} dy \, p(y, u - y) \tag{1.133}$$

The calculation we have made is a special case of a general procedure, that can be described as follows: let X be a d-dimensional random variable absolutely continuous with density $p_X(\mathbf{x})$; moreover, let $g : \mathbb{R}^d \to \mathbb{R}^k$ be a Borel-measurable function.

$W = g(X)$ is a k-dimensional random variable. If $p_W(\mathbf{y})$ were the density of W, then, for any Borel set $A \subset \mathbb{R}^k$, the following equality would hold:

$$\int_{\mathbb{R}^k} d\mathbf{y}\, 1_A(\mathbf{y}) p_W(\mathbf{y}) = P(W \in A) = P(X \in g^{-1}(A)) = \int_{\mathbb{R}^d} d\mathbf{x}\, 1_A(g(\mathbf{x})) p_X(\mathbf{x})$$
(1.134)

where $g^{-1}(A) = \{\mathbf{x} \in \mathbb{R}^d : g(\mathbf{x}) \in A\}$. The above relation is very general and, in some cases, allows to evaluate the density $p_W(\mathbf{y})$. Let's consider the case $d = k$; we assume that p_X is null outside an open set $D \subset \mathbb{R}^d$ and that $g : D \to V$ is a diffeomorphism between D and an open set $V \subset \mathbb{R}^d$. Naturally $p_W(\mathbf{y})$ will be zero outside V. Then, if $A \subset V$, a basic theorem from mathematical analysis guarantees the validity of the following change of variables:

$$\int_D d\mathbf{x}\, 1_A(g(\mathbf{x})) p_X(\mathbf{x}) = \int_V d\mathbf{y}\, 1_A(\mathbf{y}) p_X(g^{-1}(\mathbf{y})) \left| \det(J_{g^{-1}}(\mathbf{y})) \right|$$
(1.135)

which implies the following expression for the density of $W = g(X)$:

$$p_W(\mathbf{y}) = p_X(g^{-1}(\mathbf{y})) \left| \det(J_{g^{-1}}(\mathbf{y})) \right|$$
(1.136)

where $J_{g^{-1}}(\mathbf{y})$ is the Jacobian matrix of g^{-1} evaluated in $\mathbf{y}$.

1.8 Characteristic Functions

Let X be a d-dimensional random variable, $X = (X_1, \ldots, X_d)$, $X \sim \mu$.

Definition 1.11 The function ϕ_X:

$$\phi_X : \mathbb{R}^d \to \mathbb{C}$$
(1.137)

$$\theta \in \mathbb{R}^d \to \phi_X(\boldsymbol{\theta}) \overset{def}{=} E\left[\exp\left(i \sum_{k=1}^d \theta_k X_k \right) \right]$$
(1.138)

is called **characteristic function** of X.

The complex-valued integral (1.138) is defined as:

$$E\left[\exp\left(i \sum_{k=1}^d \theta_k X_k \right) \right] = E\left[\cos\left(i \sum_{k=1}^d \theta_k X_k \right) \right] + i E\left[\sin\left(\sum_{k=1}^d \theta_k X_k \right) \right]$$
(1.139)

Since trigonometric functions are measurable and limited, the abstract integrals always exist, implying that the characteristic function is well defined for every random variable. Moreover, applying the theorem of integration with respect to the law of X, we obtain:

$$\phi_X(\boldsymbol{\theta}) = \int_{\mathbb{R}^d} \exp(i\boldsymbol{\theta} \cdot \mathbf{x}) \, \mu(d\mathbf{x}) \tag{1.140}$$

which, if X is absolutely continuous, becomes:

$$\phi_X(\boldsymbol{\theta}) = \int_{\mathbb{R}^d} d\mathbf{x} \, \exp(i\boldsymbol{\theta} \cdot \mathbf{x}) \, p(\mathbf{x}) \tag{1.141}$$

If X is discrete, on the other hand, we have:

$$\phi_X(\boldsymbol{\theta}) = \sum_{\mathbf{x} \in \mathbb{R}^d} \exp(i\boldsymbol{\theta} \cdot \mathbf{x}) \, p(\mathbf{x}) \tag{1.142}$$

There is a one to one correspondence between characteristic functions and laws of random variables: $\phi_X(\boldsymbol{\theta})$ uniquely determines the law of X.

We present now some properties of characteristic functions. The first trivial observation is the equality:

$$\phi_X(0) = 1 \tag{1.143}$$

Moreover, extending to complex valued functions an inequality from abstract integration theory, we have:

$$|\phi_X(\theta)| \leq E\left[\left| \exp\left(i \sum_{k=1}^d \theta_k X_k \right) \right| \right] = 1, \forall \theta \in \mathbb{R}^d \tag{1.144}$$

that is the characteristic function is limited. Moreover, the constant function 1 (which is integrable!), dominates any sequence of functions $\exp(i\boldsymbol{\theta}_n \cdot \mathbf{x})$, in the sense of dominated convergence theorem; if $\boldsymbol{\theta}_n \to \boldsymbol{\theta}$ for $n \to +\infty$, then $\exp(i\boldsymbol{\theta}_n \cdot \mathbf{x}) \to \exp(i\boldsymbol{\theta} \cdot \mathbf{x})$ and, by dominated convergence theorem, $\phi_X(\boldsymbol{\theta}_n) \to \phi_X(\boldsymbol{\theta})$. Thus the characteristic function is continuous over $\mathbb{R}^d$.

Let's turn to smoothness. We state the following theorem, which is a plain application of dominated convergence theorem in the realm of integration theory [7]:

Theorem 1.9 *If $E[|X|^m] < +\infty$ for some integer m, then the characteristic function of X has continuous partial derivatives till order m and the following equality holds:*

$$\frac{\partial^m}{\partial \theta_{j_1} \dots \partial \theta_{j_m}} \phi_X(\boldsymbol{\theta}) = i^m \, E\left[X_{j_1} \dots X_{j_m} \exp\left(i \sum_{k=1}^d \theta_k X_k \right) \right] \tag{1.145}$$

In the case of real random variables it follows that, if $X \in L^1(\Omega, \mathcal{F}, P)$:

$$E[X] = -i\frac{d\phi_X(0)}{d\theta} \tag{1.146}$$

and, if $X \in L^2(\Omega, \mathcal{F}, P)$, we have also:

$$E[X^2] = -\frac{d^2\phi_X(0)}{d\theta^2} \tag{1.147}$$

We now present some examples:

Example 1.4 If X is **uniform** in (0, 1) we have:

$$\phi_X(\theta) = \int_0^1 dx\, e^{i\theta x} = \begin{cases} \frac{e^{i\theta}-1}{i\theta}, & \theta \neq 0 \\ 1, & \theta = 0 \end{cases} \tag{1.148}$$

Example 1.5 If X is **standard normal** we have:

$$\phi_X(\theta) = \frac{1}{\sqrt{2\pi}} \int_{-\infty}^{+\infty} dx\, \exp\left(i\theta x - \frac{x^2}{2}\right) \tag{1.149}$$

Since the density is an even function, we can write:

$$\phi_X(\theta) = \frac{1}{\sqrt{2\pi}} \int_{-\infty}^{+\infty} dx\, \cos(\theta x) \exp\left(-\frac{x^2}{2}\right) \tag{1.150}$$

We know that X is integrable and square-integrable, and thus we can apply the above theorem to evaluate the derivative of the characteristic function:

$$\frac{d\phi_X(\theta)}{d\theta} = -\frac{1}{\sqrt{2\pi}} \int_{-\infty}^{+\infty} dx\, x \sin(\theta x) \exp\left(-\frac{x^2}{2}\right) \tag{1.151}$$

We observe that the identity:

$$-\int_{-\infty}^{+\infty} dx\, x \sin(\theta x) \exp\left(-\frac{x^2}{2}\right) = \int_{-\infty}^{+\infty} dx\, \sin(\theta x) \left(\frac{d}{dx}\exp\left(-\frac{x^2}{2}\right)\right) \tag{1.152}$$

allows us to perform integration by part, providing the following result:

$$\frac{d\phi_X(\theta)}{d\theta} = -\frac{1}{\sqrt{2\pi}} \int_{-\infty}^{+\infty} dx\, \theta \cos(\theta x) \exp\left(-\frac{x^2}{2}\right) = -\theta\phi_X(\theta) \tag{1.153}$$

We have tus obtained an ordinary differential equation which, together with the initial condition $\phi_X(0) = 1$, has the unique solution:

$$\phi_X(\theta) = \exp\left(-\frac{\theta^2}{2}\right) \tag{1.154}$$

Example 1.6 If X is binomial, $X \sim B(n, p)$, we have:

$$\phi_X(\theta) = \sum_{x=0}^{n} e^{i\theta x} \binom{n}{x} p^x (1-p)^{n-x} = \left(pe^{i\theta} + 1 - p\right)^n \tag{1.155}$$

where Newton's binomial theorem has been employed.

Example 1.7 If X is Poisson with parameter λ, then:

$$\phi_X(\theta) = \sum_{x=0}^{+\infty} e^{i\theta x} \frac{\lambda^x}{x!} e^{-\lambda} = e^{-\lambda} e^{e^{i\theta}\lambda} = \exp(\lambda(e^{i\theta} - 1)) \tag{1.156}$$

We turn to independence. If $X = (Y, Z)$ is a 2-dimensional random variable absolutely continuous, with Y and Z **independent**, then, writing $\boldsymbol{\theta} = (\theta_1, \theta_2)$, we have:

$$\phi_X(\theta_1, \theta_2) = \int_{\mathbb{R}^2} dy\,dz\, e^{i\theta_1 y + \theta_2 z}\, p_Y(y) p_Z(z) = \phi_Y(\theta_1)\phi_Z(\theta_2) \tag{1.157}$$

Naturally the above equality can be trivially extended to d-dimensional random variables. In particular, if Y and Z are **independent**, we have:

$$\phi_{Y+Z}(\theta) = \phi_Y(\theta)\phi_Z(\theta) \tag{1.158}$$

It is possible to prove that, if the equality:

$$\phi_X(\theta_1, \theta_2) = \phi_Y(\theta_1)\phi_Z(\theta_2) \tag{1.159}$$

holds over the whole $\mathbb{R}^2$, then Y and Z are independent.

1.8.1   Moments and Cumulants

The characteristic function is related to the important notions of moments and cumulants, which we now introduce.

Let X be a real-valued random variable. The **moments** of X, $\{m_k\}_k$, are defined as follows:

$$m_k = E\left[X^k\right], \quad k = 0, 1, 2, \ldots \tag{1.160}$$

The **cumulants** of X, $\{\mathcal{K}_n\}_n$ are defined by the following identity:

$$\log\left(\phi_X(\theta)\right) = \sum_{n=1}^{+\infty} \mathcal{K}_n \frac{(i\theta)^n}{n!} \tag{1.161}$$

It is useful to express the first cumulants, $\mathcal{K}_1, \mathcal{K}_2, \mathcal{K}_3$, in terms of the first three moments.

To do this, we proceed formally, starting from the definition given above, written in the form:

$$E\left[\exp(tX)\right] = \exp\left(\sum_{n=1}^{+\infty} \mathcal{K}_n \frac{t^n}{n!}\right) \tag{1.162}$$

where $t = i\theta$, and Taylor expanding both sizes up to t^3:

$$1 + m_1 t + \frac{1}{2!}m_2 t^2 + \frac{1}{3!}m_3 t^3 + \cdots =$$
$$1 + \left(\mathcal{K}_1 t + \frac{1}{2!}\mathcal{K}_2 t^2 + \frac{1}{3!}\mathcal{K}_3 t^3 + \ldots\right) + \frac{1}{2!}\left(\mathcal{K}_1^2 t^2 + \mathcal{K}_1\mathcal{K}_2 t^3 + \ldots\right) + \frac{1}{3!}\left(\mathcal{K}_1^3 t^3 + \ldots\right) \tag{1.163}$$

and comparing the terms corresponding to the same power of t. We obtain:

$$\mathcal{K}_1 = m_1 = E[X], \quad \mathcal{K}_2 = m_2 - m_1^2 = Var(X) \tag{1.164}$$

and, finally:

$$\mathcal{K}_3 = m_3 - 3m_1 m_2 + 2m_1^3 \tag{1.165}$$

1.9 Normal Laws

We have already defined the standard normal law $N(0, 1)$, related to the density:

$$p(x) = \frac{1}{\sqrt{2\pi}} \exp\left(-\frac{x^2}{2}\right) \tag{1.166}$$

We have also evaluated the characteristic function of a random variable $X \sim N(0, 1)$:

$$\phi_X(\theta) = \exp\left(-\frac{\theta^2}{2}\right) \tag{1.167}$$

Moreover, we have presented the law $N(\mu, \sigma^2)$, related to the density:

$$p(x) = \frac{1}{\sqrt{2\pi}\sigma} \exp\left(-\frac{(x-\mu)^2}{2\sigma^2}\right) \tag{1.168}$$

We already know that, if $X \sim N(0, 1)$, then $Y = \sigma X + \mu \sim N(\mu, \sigma^2)$. The characteristic function of Y can thus be readily evaluated:

$$\phi_Y(\theta) = E\left[e^{i\theta Y}\right] = E\left[e^{i\theta(\sigma X + \mu)}\right] = e^{i\theta\mu}\phi_X(\sigma\theta) = \exp\left(i\theta\mu - \frac{\sigma^2\theta^2}{2}\right) \quad (1.169)$$

and the expected value and the variance are:

$$E[Y] = \mu, \quad Var(X) = \sigma^2 \quad (1.170)$$

It is useful to extend the definition of a normal law to the d-dimensional case:

Definition 1.12 We say that a d-dimensional random variable X is **normal** if its characteristic function has the form:

$$\phi_X(\boldsymbol{\theta}) = \exp\left(i\boldsymbol{\theta} \cdot \boldsymbol{\mu} - \frac{1}{2}\boldsymbol{\theta} \cdot \mathcal{C}\boldsymbol{\theta}\right) \quad (1.171)$$

where $\boldsymbol{\mu} \in \mathbb{R}^d$ and $\mathcal{C}$ is a **symmetric, positive semidefinite** real $d \times d$-matrix. We will write $X \sim N(\boldsymbol{\mu}, \mathcal{C})$.

From linear algebra we learn that, whenever $\mathcal{C}$ is a a **symmetric, positive semidefinite** real $d \times d$-matrix, there exists a **symmetric** real $d \times d$-matrix $\mathcal{A}$ such that:

$$\mathcal{A}^2 = \mathcal{C} \quad (1.172)$$

Now, if $Z = (Z_1, \ldots, Z_d)$ is a random variable such that $Z_i \sim N(0, 1), i = 1, \ldots, d$ and the Z_i are **independent**, we have:

$$\phi_Z(\boldsymbol{\theta}) = \prod_{i=1}^{d}\exp\left(-\frac{\theta_i^2}{2}\right) = \exp\left(-\frac{|\boldsymbol{\theta}|^2}{2}\right) \quad (1.173)$$

Let's define:
$$X = \mathcal{A}Z + \boldsymbol{\mu} \quad (1.174)$$

We have:

$$\phi_X(\boldsymbol{\theta}) = E[e^{i\boldsymbol{\theta}\cdot X}] = e^{i\boldsymbol{\theta}\cdot\boldsymbol{\mu}}\phi_Z(\mathcal{A}^T\boldsymbol{\theta}) = \exp\left(i\boldsymbol{\theta}\cdot\boldsymbol{\mu} - \frac{1}{2}\boldsymbol{\theta}\cdot\mathcal{C}\boldsymbol{\theta}\right) \quad (1.175)$$

where we have used the fact that $\mathcal{A}$ is symmetric and that $\mathcal{A}^2 = \mathcal{C}$. We have thus shown that, for any choice of the vector $\boldsymbol{\mu} \in \mathbb{R}^d$ and of the real, symmetric and positive semidefinite matrix $\mathcal{C}$, there exists a random variable $X \sim N(\boldsymbol{\mu}, \mathcal{C})$. Now, the random variable Z has density:

$$p_Z(\mathbf{z}) = \frac{1}{(2\pi)^{d/2}} \exp\left(-\frac{|\mathbf{z}|^2}{2}\right) \tag{1.176}$$

and it is straightforward to check that:

$$E[Z_i] = 0, \quad Cov(Z_i Z_j) = \delta_{ij} \tag{1.177}$$

Hence follows that:

$$E[X_i] = \sum_{j=1}^{d} \mathcal{A}_{ij} E[Z_j] + \mu_i = \mu_i \tag{1.178}$$

e:

$$Cov(X_i X_j) = E[(X_i - \mu_i)(X_j - \mu_j)] = \tag{1.179}$$
$$= E\left[(\sum_{k=1}^{d} \mathcal{A}_{ik} Z_k)(\sum_{l=1}^{d} \mathcal{A}_{jl} Z_l)\right] = \sum_{k,l=1}^{d} \mathcal{A}_{ik} \mathcal{A}_{jl} \delta_{kl} =$$
$$= \mathcal{C}_{ij}$$

where we have used the symmetry of $\mathcal{A}$.

If $\mathcal{C}$ is **invertible**, and thus **positive definite**, than also $\mathcal{A}$ is positive definite and X has density which is given by:

$$p_X(\mathbf{x}) = \frac{1}{|\det(\mathcal{A})|} p_Z(\mathcal{A}^{-1}(\mathbf{x} - \boldsymbol{\mu})) \tag{1.180}$$

that is:

$$p_X(\mathbf{x}) = \frac{1}{(2\pi)^{d/2}\sqrt{\det(\mathcal{C})}} \exp\left(-\frac{1}{2} \sum_{i,j=1}^{d} (x_i - \mu_i)\mathcal{C}_{ij}^{-1}(x_j - \mu_j)\right) \tag{1.181}$$

We conclude this paragraph with some important observations about normal laws. The first is that linear-affine transformations map normal random variables into normal random variables. In fact, if $X \sim N(\boldsymbol{\mu}, \mathcal{C})$ and $Y = \mathcal{B}X + \mathbf{d}$ we have:

$$\phi_Y(\boldsymbol{\theta}) = E[\exp(i\boldsymbol{\theta} \cdot Y)] = \exp(i\boldsymbol{\theta} \cdot \mathbf{d})\phi_X(^T\mathcal{B}\boldsymbol{\theta}) = \tag{1.182}$$
$$= \exp(i\boldsymbol{\theta} \cdot \mathbf{d}) \exp\left(i(^T\mathcal{B})\boldsymbol{\theta} \cdot \boldsymbol{\mu} - \tfrac{1}{2}(^T\mathcal{B})\boldsymbol{\theta} \cdot \mathcal{C}(^T\mathcal{B})\boldsymbol{\theta}\right) =$$
$$= \exp(i\boldsymbol{\theta} \cdot (\mathbf{d} + \mathcal{B}\boldsymbol{\mu}) - \tfrac{1}{2}(^T\mathcal{B})\boldsymbol{\theta} \cdot \mathcal{C}(^T\mathcal{B})\boldsymbol{\theta}) =$$
$$= \exp(i\boldsymbol{\theta} \cdot (\mathbf{d} + \mathcal{B}\boldsymbol{\mu}) - \tfrac{1}{2}\boldsymbol{\theta} \cdot (\mathcal{B}\,\mathcal{C}\,(^T\mathcal{B}))\boldsymbol{\theta})$$

that is $Y \sim N\left(\mathbf{d} + \mathcal{B}\boldsymbol{\mu}, \mathcal{B}\,\mathcal{C}\,(^T\mathcal{B})\right)$.

We consider now a real random variable of the form $Y = \mathbf{a} \cdot X = \sum_{i=1}^{d} a_i X_i$, where $\mathbf{a} \in \mathbb{R}^d$ is a vector. The following calculation:

$$\phi_Y(\theta) = E[\exp(i\theta Y)] = E[\exp(\theta \mathbf{a} \cdot X)] = \phi_X(\theta \mathbf{a}) = \qquad (1.183)$$
$$= \exp\left(i\theta \mathbf{a} \cdot \boldsymbol{\mu} - \tfrac{\theta^2}{2}\mathbf{a} \cdot \mathcal{C}\mathbf{a}\right)$$

show that $Y \sim N(\mathbf{a} \cdot \boldsymbol{\mu}, \mathbf{a} \cdot \mathcal{C}\mathbf{a})$. In particular the components of a normal are normal.

Another important observation concerns independence and non correlation. If $X_1, \ldots, X_n$ are **independent** real random variables, $X_i \sim N(\mu_i, \sigma_i^2)$, then the multi-dimensional random variable $X = (X_1, \ldots, X_n)$ is normal, $X \sim N(\boldsymbol{\mu}, \mathcal{C})$ with $\mathcal{C}_{ij} = \sigma_i^2 \delta_{ij}$, as one can trivially check writing the characteristic function. On the other hand, if $X = (X_1, \ldots, X_n)$ is normal, $X \sim N(\boldsymbol{\mu}, \mathcal{C})$, with diagonal covariance matrix $\mathcal{C}_{ij} = \sigma_i^2 \delta_{ij}$, then $X_1, \ldots, X_n$ are **independent** real random variables, $X_i \sim N(\mu_i, \sigma_i^2)$, since the characteristic function is factorized. Thus, if the joint law is normal, independence and non correlation are equivalent properties.

1.10 Convergence of Random Variables

Before introducing the very important law of large numbers and the celebrated central limit theorem, the cornerstone of probability and statistics, we summarize the basic definitions of convergence of random variables.

Definition 1.13 (*Almost sure convergence*) A sequence $\{X_n\}_{n=0}^{\infty}$ of d-dimensional random variables converges almost surely to a d-dimensional random variable X if:

$$P\left(\omega : \lim_{n\to\infty} X_n(\omega) = X(\omega)\right) = 1 \qquad (1.184)$$

Definition 1.14 (*convergence in probability*) A sequence $\{X_n\}_{n=0}^{\infty}$ of d-dimensional random variables converges in probability to a d-dimensional random variable X if:

$$\lim_{n\to\infty} P\left(\omega : |X_n(\omega) - X(\omega)| > \epsilon\right) = 0 \qquad (1.185)$$

for all $\epsilon > 0$.

Definition 1.15 (*weak convergence or convergence in law or convergence in distribution*) A sequence $\{X_n\}_{n=0}^{\infty}$ of d-dimensional random variables converges in law or weakly or in distribution to a d-dimensional random variable X if:

$$\lim_{n\to\infty} \phi_{X_n}(\theta) = \phi_X(\theta) \qquad (1.186)$$

where $\phi_{X_n}(\theta)$ and $\phi_X(\theta)$ are the characteristic functions of X_n and X respectively.

We proof now the following useful result:

Theorem 1.10 *Almost sure convergence implies convergence in probability; convergence in probability implies convergence in law.*

Proof Let's fix $\epsilon > 0$ and consider the sequence of events $\{A_n\}_{n=0}^{\infty}$:

$$A_n = \bigcup_{m=n}^{\infty} \{\omega : |X_n(\omega) - X(\omega)| > \epsilon\} \tag{1.187}$$

Such sequence is decreasing, that is $A_1 \supset A_2 \supset \dots$. We let $A_\infty = \bigcap_{n=0}^{\infty} A_n$. From the very definition of probability measure it follows that: $P(A_\infty) = \lim_{n \to +\infty} P(A_n)$.

Since the sequence is decreasing, $A_\infty = \lim_{n \to \infty} A_n = \bigcap_{n=0}^{\infty} A_n$. Moreover we have:

$$P(|X_n - X| > \epsilon) \le P(A_n) \to \lim_{n \to \infty} P(|X_n - X| > \epsilon) \le P(A_\infty) \tag{1.188}$$

It is simple to realize that, for any $\omega \in A_\infty$, $\lim_{n \to \infty} X_n(\omega) \ne X(\omega)$, and thus:

$$A_\infty \subseteq \{\omega : \lim_{n \to \infty} X_n(\omega) \ne X(\omega)\} \tag{1.189}$$

If $\{X_n\}_{n=0}^{\infty}$ converges almost surely to X, the r.h.s. of the above inclusion relation has zero probability, and thus also A_∞ has zero probability. We have thus:

$$\lim_{n \to \infty} P(|X_n - X| > \epsilon) = 0 \tag{1.190}$$

and the first part of the theorem is proved.

Let's consider now:

$$\left|\phi_{X_n}(\theta) - \phi_X(\theta)\right| = \left|E[e^{i\theta X_n}] - E[e^{i\theta X}]\right| \le E\left[\left|e^{i\theta(X_n - X)} - 1\right|\right] \tag{1.191}$$

Since the function $x \to e^{i\theta x}$ is continuous, for all $\epsilon > 0$ there exists $\delta > 0$ such that $|x| < \delta$ and $|e^{i\theta x} - 1| < \epsilon$. We can rewrite the last member of the above inequality as:

$$E\left[\left|e^{i\theta(X_n - X)} - 1\right|\Theta(|X_n - X| < \delta)\right] + E\left[\left|e^{i\theta(X_n - X)} - 1\right|\Theta(|X_n - X| \ge \delta)\right] \tag{1.192}$$

so that the following inequality is at hand:

$$\left|\phi_{X_n}(\theta) - \phi_X(\theta)\right| \le \epsilon + 2\,P(|X_n - X| \ge \delta) \tag{1.193}$$

So, if $\{X_n\}_{n=0}^{\infty}$ converges in probability to X:

$$\lim_{n \to \infty} \left|\phi_{X_n}(\theta) - \phi_X(\theta)\right| \le \epsilon \tag{1.194}$$

for any $\epsilon > 0$, and thus $\lim_{n \to \infty} \phi_{X_n}(\theta) = \phi_X(\theta)$.

Simple examples show that other implications do not hold in general.

1. Suppose $\{\Omega, \mathcal{F}, P\}$ be the probability space with $([0, 1], \mathcal{B}([0, 1]), \mu)$, where μ is the uniform probability. Let also:

$$
\begin{aligned}
X_1(\omega) &= \omega + \chi_{[0,1]}(\omega) \\
X_2(\omega) &= \omega + \chi_{[0,1/2]}(\omega) \\
X_3(\omega) &= \omega + \chi_{[1/2,1]}(\omega) \\
X_4(\omega) &= \omega + \chi_{[0,1/3]}(\omega) \\
X_5(\omega) &= \omega + \chi_{[1/3,2/3]}(\omega) \\
X_6(\omega) &= \omega + \chi_{[2/3,1]}(\omega) \\
&\cdots \\
X_n(\omega) &= \omega + \chi_{I_n}(\omega)
\end{aligned}
\tag{1.195}
$$

and $X(\omega) = \omega$. Since:

$$
P(\omega : |X_n(\omega) - X(\omega)| > \epsilon) = P(\omega \in I_n)
\tag{1.196}
$$

the sequence $\{X_n\}_{n=0}^{\infty}$ converges in probability to X. Nevertheless, since $\lim_{n \to \infty} X_n(\omega)$ does not exist for any $\omega \in \Omega$, it does not converge almost surely to X.

2. Suppose $\{\Omega, \mathcal{F}, P\}$ be the probability space with $(\{0, 1\}, \mathcal{P}(\{0, 1\}), \mu)$, where μ is the uniform probability. Let also:

$$
X_n(\omega) = \omega
\tag{1.197}
$$

and $X(\omega) = 1 - \omega$. Since:

$$
\phi_{X_n}(\theta) = E[e^{i\theta X_n}] = \frac{1}{2}e^{i\theta} \quad \phi_X(\theta) = E[e^{i\theta X}] = \frac{1}{2}e^{i\theta}
\tag{1.198}
$$

the sequence $\{X_n\}_{n=0}^{\infty}$ converges in law to X. On the other hand, since $X_n - X = 1$, it does not converge in probability to X.

1.11 The Law of Large Numbers and the Central Limit Theorem

We conclude our review of probability theory with two key results about convergence and approximation.

Let $\{X_k\}_{k \geq 1}$ be a sequence of real valued random variables, **independent** and **identically distributed (iid)**: this means that all the X_k have the same law. We also assume that all the X_k are square-integrable, and we introduce the notation:

$$
\mu = E[X_k], \quad \sigma^2 = Var(X_k)
\tag{1.199}
$$

μ and σ^2 are finite by construction, and do not depend on k because we have assumed the random variables identically distributed. Let's define the **empirical mean**:

$$S_n = \frac{1}{n} \sum_{k=1}^{n} X_n \qquad (1.200)$$

We perform now some calculations:

$$E[S_n] = \frac{1}{n} \sum_{k=1}^{n} E[X_n] = \mu \ , \qquad (1.201)$$

$$
\begin{aligned}
Var(S_n) &= E[(S_n - \mu)^2] = E[S_n^2] - \mu^2 = \qquad (1.202)\\
&= \tfrac{1}{n^2} \sum_{i,j=1}^{n} E[X_i X_j] - \mu^2 = \\
&= \tfrac{1}{n^2} \sum_{i=1}^{n} E[X_i^2] + \tfrac{1}{n^2} \sum_{i \neq j=1}^{n} E[X_i X_j] - \mu^2 = \\
&= \tfrac{1}{n^2} \sum_{i=1}^{n} \left(Var(X_i) + E[X_i]^2 \right) + \tfrac{1}{n^2} \sum_{i \neq j=1}^{n} E[X_i] E[X_j] - \mu^2 = \\
&= \tfrac{\sigma^2}{n} + \tfrac{\mu^2}{n} + \tfrac{n(n-1)}{n^2} \mu^2 - \mu^2 = \tfrac{\sigma^2}{n}
\end{aligned}
$$

We use now the Chebyshev inequality:

$$P\left(|S_n - \mu| \geq \eta\right) \leq \frac{Var(S_n)}{\eta^2} = \frac{\sigma^2}{n\eta^2} \xrightarrow{n \to +\infty} 0 \qquad (1.203)$$

We have proved in this way a very important result:

Theorem 1.11 (Weak law of large numbers) *The sequence of empirical means* $\{S_n\}_{n \geq 1}$ *of independent and indentically distributed real square-integrable random variables* $\{X_k\}_{k \geq 1}$ *with expected value* μ, **converges in probability** *to* μ:

$$\lim_{n \to +\infty} P\left(|S_n - \mu| \geq \eta\right) = 0, \quad \forall \eta > 0 \qquad (1.204)$$

It can be shown that this convergence result can be proved also under weakened hypotheses, removing the assumption of finite variance, and with a stronger notion of convergence:

Theorem 1.12 (Strong law of large numbers) *The sequence of empirical means* $\{S_n\}_{n \geq 1}$ *of independent and indentically distributed real integrable random variables* $\{X_k\}_{k \geq 1}$ *with expected value* μ, **converges almost surely** *to* μ.

We omit the proof of such result.

Now, let's introduce:

$$S_n^\star = \frac{S_n - \mu}{\sigma/\sqrt{n}} \tag{1.205}$$

We may write:

$$S_n^\star = \frac{1}{\sqrt{n}} \sum_{k=1}^{n} Y_k, \quad Y_k = \frac{X_k - \mu}{\sigma} \tag{1.206}$$

where the random variables Y_k are clearly independent and identically distributed, and satisfy:

$$E[Y_k] = 0, \quad Var(Y_k) = 1 \tag{1.207}$$

We observe that the expression:

$$S_n^\star = \frac{Y_1}{\sqrt{n}} + \cdots + \frac{Y_n}{\sqrt{n}} \tag{1.208}$$

suggests the idea of a sum of many small independent non systematic (zero mean) effects. This could remind the reader the theory of errors which he/she has learned in university courses.

Since the Y_k are identically distributed, they have the same characteristic function, which we will denote simply ϕ. We evaluate now the characteristic function of $S_n^\star$:

$$\phi_{S_n^\star}(\theta) = E[\exp(i\theta S_n^\star)] = E[\exp(i\theta \tfrac{1}{\sqrt{n}} \sum_{k=1}^{n} Y_k)] = \tag{1.209}$$

$$= \left(\phi(\tfrac{\theta}{\sqrt{n}})\right)^n = \exp\left(n \log(\phi(\tfrac{\theta}{\sqrt{n}}))\right) = \exp\left(n \log\left(1 + \phi(\tfrac{\theta}{\sqrt{n}}) - 1\right)\right)$$

Since the characteristic functions are always continuous, we have $\phi(\tfrac{\theta}{\sqrt{n}}) \to \phi(0) = 1$ if $n \to +\infty$ for any fixed θ; this implies the asymptotic behavior:

$$n \log\left(1 + \phi(\tfrac{\theta}{\sqrt{n}}) - 1\right) \overset{n \to +\infty}{\sim} n\left(\phi(\tfrac{\theta}{\sqrt{n}}) - 1\right) \tag{1.210}$$

The Y_k are by construction square-integrable, and thus:

$$\frac{d\phi(0)}{d\theta} = i E[Y_k] = 0, \quad \frac{d^2\phi(0)}{d\theta^2} = -Var(Y_k) = -1 \tag{1.211}$$

so that:

$$\phi(\tfrac{\theta}{\sqrt{n}}) - 1 = -\frac{\theta^2}{2n} + o(\tfrac{1}{n}) \tag{1.212}$$

which implies:

$$n \log \left(1 + \phi(\frac{\theta}{\sqrt{n}}) - 1\right) \overset{n \to +\infty}{\sim} n \left(\phi(\frac{\theta}{\sqrt{n}}) - 1\right) \overset{n \to +\infty}{\sim} -\frac{\theta^2}{2} \qquad (1.213)$$

We have thus found the following very important result:

$$\lim_{n \to +\infty} \phi_{S_n^\star}(\theta) = \exp\left(-\frac{\theta^2}{2}\right) \qquad (1.214)$$

where in the right hand side we have the characteristic function of a standard normal random variable $Z \sim N(0, 1)$.

We summarize what we have found in the following central result in probability theory, the cornerstone of mathematical statistics:

Theorem 1.13 (Central Limit theorem) *If* $\{X_k\}_{k \geq 1}$ *is a sequence of real valued square integrable random variables,* **independent** *and* **identically distributed**, *letting* $\mu = E[X_k]$ *and* $\sigma^2 = Var(X_k)$, *the sequence:*

$$S_n^\star = \frac{\frac{1}{n}\sum_{k=1}^{n} X_k - \mu}{\sigma / \sqrt{n}} \qquad (1.215)$$

converges in law *to a standard normal random variable* $Z \sim N(0, 1)$.

We use the notation:

$$S_n^\star \overset{D}{\to} Z \sim N(0, 1), \quad n \to +\infty \qquad (1.216)$$

where the arrow indicate convergence in law.

1.12   Further Readings

This chapter contains all the notions that are needed to understand the remainder of the book, but it does not aim to provide a thorough exposition of probability theory. Readers wishing to deepen their knowledge about basic probability theory can refer to many excellent introductory textbooks, like, e.g., [2–6].

Readers interested in abstract integration theory can in turn refer, for example, to [7].

Finally, for application to statistical and quantum mechanics, possible further readings are [8, 9].

Problems

1.1 Expected values and variances
Evaluate $E[X]$ and $Var(X)$ when:

1. X is uniform in $(0, 1)$.

2. X is standard normal, $X \sim N(0, 1)$.

3. X is binomial, $X \sim B(n, p)$.

4. X is Poisson with parameter λ.

1.2 Useful formulas
Let T be a random variable taking values in $\mathbb{N} = \{1, 2, 3, \ldots, \}$. Prove that

$$E[T] = \sum_{n=1}^{\infty} P(T \geq n) \tag{1.217}$$

Let X be an absolutely continuous random variable, such that $X \geq 0$ and suppose $F(x)$ is its cumulative distribution function. Prove that:

$$E[X] = \int_0^{+\infty} dy \, P(X \geq y) = \int_0^{+\infty} dy \, (1 - F(y)) \tag{1.218}$$

1.3 Random point on a circumference
Consider a circumference Γ of unitary radius, fix a point $A \in \Gamma$ and suppose that we select another point $B \in \Gamma$, defined by an angle θ chosen with uniform distribution in $[0, 2\pi]$. What is the probability distribution of the length L_{AB} of the chord connecting A and B? What are its expectation and variance? What is the probability that L_{AB} is longer than the side $\sqrt{3}$ of an equilater triangle inscribed in Γ?

1.4 Bertrand paradox
The previous exercise is useful to introduce a very interesting historical paradox, which makes evident the importance of the choice of the probability space for the description of a phenomenon. Suppose we face quite the same problem but with a slightly less precise formulation. Given a circumference Γ of unitary radius, suppose that a chord of the circumference is chosen randomly. What is the probability that the chord is longer than the side of an equilater triangle inscribed in the circumference $L = \sqrt{3}$? The key point is how we model the randomness of the choice of the chord.

In the previous problem we have chosen a model for describing the way we pick up the chord. Different choices lead to different results! Statistics is the discipline which can tell us which is more reliable! Evaluate the requested probability using the following other two different models:

1. Suppose that, fixed a point A on the circumference a point x is chosen in the radius ending on A with uniform probability. The random chord is the one passing through x and perpendicular to the radius ending on A.

2. Suppose that a point x is chosen randomly anywhere within the circle having Γ as its boundary. The random chord is the one having such point as its midpoint.

1.5 Gaussian integration
Verify explicitly that the probability density:

$$p_X(\mathbf{x}) = \frac{1}{(2\pi)^{d/2}\sqrt{\det(\mathcal{C})}} \exp\left(-\frac{1}{2}\sum_{i,j=1}^{d}(x_i-\mu_i)\mathcal{C}_{ij}^{-1}(x_j-\mu_j)\right) \qquad (1.219)$$

satisfies the normalization condition $\int_{\mathbb{R}^d} d\mathbf{x}\, p_X(\mathbf{x}) = 1$, whenever the covariance matrix is positive definite.

Find out an explicit expression for the gaussian integral:

$$I = \int_{\mathbb{R}^d} d\mathbf{x}\, \exp\left(-\frac{1}{2}\sum_{i,j=1}^{d} x_i\mathcal{O}_{ij}x_j + \sum_{i=1}^{d}\vartheta_i x_i\right) \qquad (1.220)$$

when the matrix $\mathcal{O}$ is real, symmetric and positive definite, and for the derivative:

$$\frac{\partial^2 I}{\partial\vartheta_i\partial\vartheta_j}\Big|_{\vartheta=0} \qquad (1.221)$$

Do you see any relation between this expression and covariance matrix of multidimensional normal random variables?

1.6 Isserlis identity and Wick theorem
Let $Z = (Z_1 \ldots Z_{2n})$ be a normal $2n$-dimensional random variable $Z \sim N(0,\mathcal{C})$. Prove Isserlis Identity:

$$E[Z_1 \ldots Z_{2n}] = \frac{1}{2^n n!}\sum_{\sigma\in S_{2n}} E[Z_{\sigma(1)}Z_{\sigma(2)}]\ldots E[Z_{\sigma(2n-1)}Z_{\sigma(2n)}] \qquad (1.222)$$

where S_{2n} contains all the $2n!$ permutations of the $2n$ labels. Rearrange the above expression to obtain the very famous Wick theorem (for bosons):

$$E[Z_1 \ldots Z_{2n}] = \sum_{\text{all pairings}} \mathcal{C}_{i_1 j_1}\ldots\mathcal{C}_{i_n j_n} \qquad (1.223)$$

where the summation involves all the $(2n)!/(2^n n!)$ possible ways to build up ordered pairs (i_l, j_l), $i_l < j_l$, $l = 1,\ldots,n$ from the set $(1,\ldots,2n)$.

1.7 $\chi^2(n)$ **random variable: chi-square law with n degrees of freedom.**
Let $X_1 \ldots X_n$ be iid (independent and identically distributed) standard normal random variables. Compute the probability density and the moments of the random variable $Y \sim \chi^2(n)$:

$$Y = X_1^2 + \ldots X_n^2 \tag{1.224}$$

1.8 $t(n)$ **random variable: Student law with n degrees of freedom.**
Let X and Y be independent random variables such that $X \sim N(0, 1)$ and $Y \sim \chi^2(n)$. Compute the probability density and the moments of the random variable $T \sim t(n)$:

$$T = \frac{X}{\sqrt{\frac{Y}{n}}} \tag{1.225}$$

References

1. Kolmogorov, A.N.: Foundations of the Theory of Probability (1933) Paperback
2. Feller, W.: An introduction to Probability Theory and its Applications. Wiley (1966)
3. Chung, K.L.: A Course in Probability Theory. Harcourt, Brace & World (1968)
4. Breiman, L.: Probability. Addison-Wesley (1992)
5. Jacod, J., & Protter, P. (2004). *Probability Essentials*. Berlin: Springer.
6. Baldi, P. (1998). *Calcolo delle probabilitá e statistica*. Milano: McGraw-Hill.
7. Rudin, W.: Real and Complex Analysis. McGraw-Hill (1987)
8. Huang, K.: Introduction to Statistical Physics. Taylor & Francis (2001)
9. Schwabl, F., Brewer, W D.: Statistical Mechanics. Springer (2006)

Chapter 2
Applications to Mathematical Statistics

Abstract In this chapter we present applications of probability theory within the science of extracting information from data: mathematical statistics. We present, on a rigorous basis, the theory of statistical estimators and some of the most widely employed hypothesis tests. Finally, we briefly discuss linear regression, a mandatory topic for physicists.

Keywords Statistical models · Estimators · Cochran theorem
Cramer-Rao theorem · Maximum likelihood estimators · Hypothesis tests · Linear regression

2.1 Statistical Models

One very important environment in which the formalism of probability theory plays a leading role is the science of extracting information from **data**: mathematical statistics. We review the key results of mathematical statistics, since they are a mandatory requisite for any scientist working on data, either coming from experiments or from numerical simulations.

Observations, that is measured data, are used to **infer** the values of some parameters necessary to complete a mathematical description of an experiment. The simplest situation we can imagine is to measure n-times the same quantity (or the same set of quantities), performing all the measurements under the same experimental conditions and in such a way that the result of any measure does not affect the results of the others. The outcome of the experiment is thus a collection of **data**:

$$(\mathbf{x}_1 \ldots \mathbf{x}_n) \tag{2.1}$$

where $\mathbf{x}_i \in \mathbb{R}^k$ is the result of the i-th measure.

Naturally, even if we are very careful in the preparation of the experimental setup, we cannot expect that, if we repeated the whole set of n measures, we would find the same data (2.1): some randomness unavoidably exists.

© Springer International Publishing AG, part of Springer Nature 2018
E. Vitali et al., *Theory and Simulation of Random Phenomena*, UNITEXT
for Physics, https://doi.org/10.1007/978-3-319-90515-0_2

It is thus natural to use the language of probability theory to describe the experiment. The i-th measure can be modeled by a random variable X_i, defined on some probability space $(\Omega, \mathcal{F}, P)$ and the whole outcome of the experiment, the data $(\mathbf{x}_1 \ldots \mathbf{x}_n)$, can be viewed as realizations of a collection $(X_1, \ldots, X_n)$ of random variables, **independent** and **identically distributed**. The requirement of independence is suggested by the assumption that the result of any measure does not affect the results of the others while the one of identical distribution translates the idea that the measurements are performed under the same experimental conditions.

Sometimes, depending on the measurement procedure, one has an idea about the law of the random variables X_i: in could be Binomial, Poisson, Exponential, Normal, Uniform and so on. However, in general, the actual parameters characterizing the law are not known but can be **inferred** from the data $(\mathbf{x}_1 \ldots \mathbf{x}_n)$.

This typical situation justifies the following definition:

Definition 2.1 A **statistical model** is a family:

$$\{(\Omega, \mathcal{F}, P_\theta)\}_{\theta \in \Theta}$$

of probability spaces sharing the same sample space and the same collection of events. The probability measure $P_\theta : \mathcal{F} \to [0, 1]$ depends on a parameter θ taking values in a set $\Theta \subseteq \mathbb{R}^m$.

A statistical model describes the preparation procedure of the experiment; the results of the experiment, as anticipated above, are realizations of a multidimensional random variable:

$$X = (X_1 \ldots X_n) \tag{2.2}$$

where the components $X_i : \Omega \to E \subset \mathbb{R}^k$ are **independent** and **identically distributed**. Such a random variable X is called a **sample** of **rank** n.

Within a statistical model, the law of X_i, describing the measurement procedure:

$$B \in \mathcal{B}(\mathbb{R}^k) \to \mu_\theta(B) = P_\theta(X_i \in B) \tag{2.3}$$

naturally depends on θ and can be related to a density $p_\theta(\mathbf{x})$, discrete of continuous.

As usual in probability theory, the actual precise definition of the triplet $(\Omega, \mathcal{F}, P_\theta)$ is in general omitted once the law of the sample is specified.

Some examples of models that are frequently employed are summarized in the following table, in which we indicate the range of the measurements E, the set Θ and the density $p_\theta(\mathbf{x})$.

All such models are examples of a wide class of models, called s-**parameters exponential models**, characterized by densities of the form:

$$p_\theta(\mathbf{x}) = e^{-\mathbf{f}_1(\theta) \cdot \mathbf{f}_2(\mathbf{x})} \, e^{-f_3(\theta)} f_4(\mathbf{x}) \tag{2.4}$$

Model	E	Θ	p_θ
Bernoulli	$\{0, 1\}$	$(0, 1)$	$\theta^{1-x}(1 - \theta)^x$
Gaussian	$\mathbb{R}$	$\mathbb{R} \times (0, \infty)$	$\frac{1}{\sqrt{2\pi\theta_1}} e^{-\frac{(x-\theta_0)^2}{2\theta_1}}$
Exponential	$(0, \infty)$	$(0, \infty)$	$\theta_0 \, e^{-\theta_0 x}$

where $\mathbf{f}_1 : \Theta \to \mathbb{R}^s, \mathbf{f}_2 : \mathbb{R}^k \to \mathbb{R}^s, f_3 : \Theta \to \mathbb{R}$ ed $f_4 : \mathbb{R}^k \to [0, \infty)$.

2.2 Estimators

One of the main goals of mathematical statistics is to use the **data** $(\mathbf{x}_1 \ldots \mathbf{x}_n)$ to **estimate** functions $\tau(\theta)$ of the parameter θ, useful to complete the probabilistic description of the experiment. For this purpose, suitable functions have to be applied to the data; keeping in mind that the data are viewed as realizations of a sample, the following definition is quite natural:

Definition 2.2 A **statistic** $\mathcal{T}$ is an s-dimensional random variable of the form:

$$\omega \in \Omega \to \mathcal{T}(\omega) = t(X_1(\omega), \ldots, X_n(\omega))$$

where $t : \mathbb{R}^k \times \cdots \times \mathbb{R}^k \to \mathbb{R}^s$ is a measurable function which **does not depend** on the parameter θ and $(X_1 \ldots X_n)$ is a sample of rank n.

The Definition (2.2) of a statistic describes the manipulations we make to the data. When a statistic $\mathcal{T}$ is used to infer a value for a given function of the parameter θ, $\tau(\theta)$, we say that $\mathcal{T}$ is an **estimator** of $\tau(\theta)$, while $t(\mathbf{x}_1 \ldots \mathbf{x}_n)$ is called **pointwise estimation** of $\tau(\theta)$.

2.3 The Empiric Mean

The most natural statistic one considers when dealing with a set of data is the **mean**. Inside our formalism, we build the estimator $\mathcal{M}$:

$$\mathcal{M} = m(X_1 \ldots X_n) = \frac{\sum_{i=1}^n X_i}{n} \tag{2.5}$$

for the unknown quantity:

$$\mu(\boldsymbol{\theta}) = E_{\boldsymbol{\theta}}(X_i) = \int dx\, x\, p_{\boldsymbol{\theta}}(x)$$

We start our presentation of mathematical statistics from the analysis of this estimator, which will help us to introduce some basic notions.

Intuitively, given the set of data $(x_1, \ldots, x_n)$, one would like to write something like $\mu(\boldsymbol{\theta}) \simeq \frac{\sum_{i=1}^{n} x_i}{n}$.

Let's give a precise meaning to such an operative procedure.

$\mathcal{M}$ is a random variable, with a law depending on the law of the sample; in particular the expected value is readily computed:

$$E_{\boldsymbol{\theta}}(\mathcal{M}) = \frac{\sum_{i=1}^{n} E_{\boldsymbol{\theta}}(X_i)}{n} = \mu(\boldsymbol{\theta})$$

and coincides with $\mu(\boldsymbol{\theta})$. So $\mathcal{M}$ has expected value equal to the quantity we wish to infer. This is an important property of the estimator, called **unbiasedness**, defined in the following:

Definition 2.3 An estimator $\mathcal{T}$ of a function $\tau(\boldsymbol{\theta})$ is called **unbiased** if:

$$E_{\boldsymbol{\theta}}(\mathcal{T}) = \tau(\boldsymbol{\theta}), \quad \forall \boldsymbol{\theta} \tag{2.6}$$

What about the "error"? In other words, what do we expect about the spreading of the realizations of $\mathcal{M}$ around the expected value $\mu(\boldsymbol{\theta})$? This is controlled by the variance of $\mathcal{M}$, which we have already computed in the chapter of probability. Letting:

$$\sigma^2(\boldsymbol{\theta}) = Var_{\boldsymbol{\theta}}(X_i) = \int dx (x - \mu(\boldsymbol{\theta}))^2 p_{\boldsymbol{\theta}}(x)$$

we have:

$$Var_{\boldsymbol{\theta}}(\mathcal{M}) = \frac{\sigma^2(\boldsymbol{\theta})}{n}$$

As we have already learnt when studying the law of large numbers, the following inequality holds:

$$P_{\boldsymbol{\theta}}\left(|\mathcal{M} - \mu(\boldsymbol{\theta})| > \eta\right) \le \frac{\sigma^2(\boldsymbol{\theta})}{n\eta^2}$$

for any $\eta > 0$. This means that, provided that the $\sigma^2(\boldsymbol{\theta}) < +\infty$ for all $\boldsymbol{\theta}$, increasing the number of data, i.e. the rank of the sample, the spreading of $\mathcal{M}$ around the expected value $\mu(\boldsymbol{\theta})$ becomes smaller and smaller, making the number $\frac{\sum_{i=1}^{n} x_i}{n}$ nearer and nearer to $\mu(\boldsymbol{\theta})$ for any realization of the sample.

This is another useful property of an estimator, **consistency**, expressed in general in the following:

Definition 2.4 An estimator $\mathcal{T}$ of a function $\tau(\theta)$ is called **consistent** if it converges in probability to $\tau(\theta)$ if the rank of the sample tends to $+\infty$.

In order to proceed further, we need some assumption about the law of the X_i. A typical situation, quite always presented in textbooks, is the case when the X_i are **normal** with **known** variance σ^2. This can be a good model if we know a priori the sensibility of a given instrument, and is given by:

$$\theta \rightarrow \mu_\theta(B) = P(X_i \in B) = \int_B dx \frac{\exp\left(-\frac{(x-\theta)^2}{2\sigma^2}\right)}{\sqrt{2\pi\sigma^2}} \tag{2.7}$$

where the parameter $\theta \equiv \mu(\theta)$ is to be inferred from the data, while σ^2 is a fixed parameter, which we assume to know a priori. In such case, $\mathcal{M}$ is normal, being a linear combination of normal random variables. In particular, we have:

$$\mathcal{Z} = \frac{\mathcal{M} - \theta}{\sqrt{\frac{\sigma^2}{n}}} \sim N(0, 1) \tag{2.8}$$

This means that the statistic $\mathcal{M}$ is normally distributed around the unknown mean θ with a known variance, decreasing with the rank of the sample.

Before proceeding, we need the very important definition of **quantiles**.

Definition 2.5 If X is a real valued absolutely continuous random variable, the quantile of order $\alpha \in (0, 1)$, q_α, is defined by the following:

$$P(X \leq q_\alpha) = \alpha \tag{2.9}$$

Introducing the **quantiles** ϕ_α of a standard normal law, we can write the exact result:

$$P_\theta\left(-\phi_{1-\frac{\alpha}{2}} \leq \mathcal{Z} \leq \phi_{1-\frac{\alpha}{2}}\right) = 1 - \alpha \tag{2.10}$$

or, equivalently:

$$P_\theta\left(\mathcal{M} - \phi_{1-\frac{\alpha}{2}}\sqrt{\frac{\sigma^2}{n}} \leq \theta \leq \mathcal{M} + \phi_{1-\frac{\alpha}{2}}\sqrt{\frac{\sigma^2}{n}}\right) = 1 - \alpha \tag{2.11}$$

It is important to understand the meaning of this equality: let's fix $1 - \alpha = 0.95 = 95\%$, the **confidence level**; in such case, we have to use the quantile $\phi_{1-\frac{\alpha}{2}} = \phi_{0.975} = 1.96$. The **random interval**:

$$\left[\mathcal{M} - 1.96\sqrt{\frac{\sigma^2}{n}}, \mathcal{M} + 1.96\sqrt{\frac{\sigma^2}{n}} \right] \tag{2.12}$$

contains the unknown expected value θ, with probability $1 - \alpha = 95\%$: for that reason, it is called **confidence interval** at the level $1 - \alpha = 95\%$ for the parameter θ. This is a particular example of the following very general definition:

Definition 2.6 Given two real valued statistics $\mathcal{A}, \mathcal{B}$, the random interval $[\mathcal{A}, \mathcal{B}]$ is called **confidence interval** at the level $1 - \alpha \in (0, 1)$ of a function $\tau(\theta)$ if:
$$P_\theta(\tau(\theta) \in [\mathcal{A}, \mathcal{B}]) \geq 1 - \alpha \quad \forall \theta \in \Theta$$

To summarize, if our data $(x_1, \ldots, x_n)$ can be modelled with normal random variables with unknown expected value θ and known variance σ^2, the real number:

$$\bar{x} = \frac{\sum_{i=1}^{n} x_i}{n} \tag{2.13}$$

is a pointwise estimation of θ. Moreover, letting:

$$\delta\bar{x} = \sqrt{\frac{\sigma^2}{n}} \tag{2.14}$$

the interval:
$$[\bar{x} - 1.96\delta\bar{x}, \bar{x} + 1.96\delta\bar{x}] \tag{2.15}$$

is an estimation of a confidence interval at the level 95% for the unknown θ.

2.4 Cochran Theorem and Estimation of the Variance

A far more general situation emerges in the case that the sample components are normal with both expectation and variance unknown.

$$\theta = (\theta_0, \theta_1) \rightarrow \mu_\theta(B) = P(X_i \in B) = \int_B dx \frac{\exp\left(-\frac{(x-\theta_0)^2}{2\theta_1}\right)}{\sqrt{2\pi\theta_1}} \tag{2.16}$$

We show now how to use the data to estimate the unknown functions $\mu(\theta)$ and $\sigma^2(\theta)$, and to obtain two intervals in $\mathbb{R}$ containing respectively the two functions with a given confidence level.

Besides the statistic $\mathcal{M}$ introduced above, which is an unbiased and consistent estimator of $\mu(\boldsymbol{\theta}) = \theta_0$, we introduce the following estimator:

$$\mathcal{S}^2 = s^2(X_1 \ldots X_n) = \frac{\sum_{i=1}^n (X_i - \mathcal{M})^2}{n-1} \tag{2.17}$$

of $\sigma^2(\boldsymbol{\theta}) = \theta_1$. The presence of $n-1$ in the denominator makes $\mathcal{S}^2$ unbiased, as follows from the following calculation:

$$E_\theta(\mathcal{S}^2) = E_\theta \left[\frac{\sum_{i=1}^n X_i^2 - n \left(\frac{\sum_{j=1}^n X_j}{n} \right)^2}{n-1} \right] =$$

$$= \frac{nE_\theta(X_i^2) - \frac{1}{n} E_\theta \left(\left(\sum_{j=1}^n X_j \right)^2 \right)}{n-1} =$$

$$= \frac{nE_\theta(X_i^2) - \frac{1}{n} \left(Var_\theta \left(\sum_{j=1}^n X_j \right) + \left(E_\theta \left(\sum_{j=1}^n X_j \right) \right)^2 \right)}{n-1} =$$

$$= \frac{nE_\theta(X_i^2) - Var_\theta(X_i) - n (E_\theta(X_i))^2}{n-1} = Var_\theta(X_i) = \sigma^2(\boldsymbol{\theta})$$

$\mathcal{S}^2$ is also consistent, as can be easily verified using the law of large numbers. The pointwise estimation of $\sigma^2(\boldsymbol{\theta})$ is thus:

$$s^2(x_1 \ldots x_n) = \frac{\sum_{i=1}^n \left(x_i - \frac{\sum_{j=1}^n x_j}{n} \right)^2}{n-1} \tag{2.18}$$

Let's turn to confidence intervals.

The central result is that the random variable:

$$\mathcal{R} = \frac{\mathcal{M} - \theta_0}{\sqrt{\frac{\mathcal{S}^2}{n}}} \tag{2.19}$$

follows a **Student law** $t(n-1)$ with $n-1$ degrees of freedom, and this allows to build confidence intervals for the mean, θ_0. Moreover, it turns out that the random variable:

$$\frac{n-1}{\theta_1} \mathcal{S}^2 \tag{2.20}$$

follows a **chi-square law** $\chi^2(n-1)$ with $n-1$ degrees of freedom, allowing to compute confidence intervals for the variance, θ_1.

We let $t_{1-\alpha/2}(n-1)$ be the **quantile** of order $1 - \alpha/2$ of the Student law $t(n-1)$, defined, as we have already seen before, by the following:

$$P_\theta(\mathcal{R} \le t_{1-\alpha/2}(n-1)) = 1 - \alpha/2 \tag{2.21}$$

Since the Student law is even ($\mathcal{R}$ and $-\mathcal{R}$ have the same law), we have $t_{\alpha/2}(n-1) = -t_{1-\alpha/2}(n-1)$ and thus we can write:

$$P_\theta(-t_{1-\alpha/2}(n-1) \le \mathcal{R} \le t_{1-\alpha/2}(n-1)) = 1 - \alpha \tag{2.22}$$

We conclude that the random interval:

$$[\mathcal{M} - t_{1-\alpha/2}(n-1)\sqrt{\frac{\mathcal{S}^2}{n}}, \mathcal{M} + t_{1-\alpha/2}(n-1)\sqrt{\frac{\mathcal{S}^2}{n}}] \tag{2.23}$$

is a **confidence interval** at the level $1 - \alpha$ for the mean $\mu(\boldsymbol{\theta}) = \theta_0$. If we replace the estimators with the pointwise estimations, we provide an estimation for the confidence interval which, letting:

$$m = \frac{\sum_{i=1}^n x_i}{n}, \quad s^2 = \frac{1}{n-1}\sum_{i=1}^n (x_i - m)^2$$

is:

$$[m - t_{1-\alpha/2}(n-1)\sqrt{\frac{s^2}{n}}, m + t_{1-\alpha/2}(n-1)\sqrt{\frac{s^2}{n}}]$$

The most typical choice is the level 95%, which means $1 - \alpha = 0.95$, that is $\alpha = 0.05$: we have to use the quantile $t_{1-\alpha/2}(n-1) = t_{0.975}(n-1)$, which, for example in the case $n = 100$, is nearly 1.985.

Remark 2.1 Quite often, when the rank of the sample in large, in the expression of confidence intervals one replaces the quantile $t_{1-\alpha/2}(n-1)$ of the Student law with the ones of standard normal law $\phi_{1-\alpha/2}$ (naturally strictly independent on n). In the case $n = 100$, for example, such substitution would give a slightly smaller interval, being $\phi_{0.975} = 1.96$.

For the variance, the result:

$$\frac{(n-1)\mathcal{S}^2}{\theta_1} \sim \chi^2(n-1) \tag{2.24}$$

implies that:

$$P_\theta\left(\chi^2_{\alpha/2}(n-1) \le \frac{(n-1)\mathcal{S}^2}{\theta_1} \le \chi^2_{1-\alpha/2}(n-1)\right) = 1 - \alpha \tag{2.25}$$

where we have introduced the quantiles of the $\chi^2(n-1)$. We can rewrite the above expression in the following way:

$$P_\theta \left(\frac{(n-1)S^2}{\chi^2_{1-\alpha/2}(n-1)} \leq \theta_1 \leq \frac{(n-1)S^2}{\chi^2_{\alpha/2}(n-1)} \right) = 1 - \alpha \qquad (2.26)$$

which shows that:

$$\left[\frac{(n-1)S^2}{\chi^2_{1-\alpha/2}(n-1)}, \frac{(n-1)S^2}{\chi^2_{\alpha/2}(n-1)} \right] \qquad (2.27)$$

is a confidence interval at the level $1 - \alpha$ for $\sigma^2(\theta) = \theta_1$.

Typically, when a sample has rank $n > 30$, the following approximation turns out to be very accurate:

$$\chi^2_{1-\alpha/2}(n-1) \simeq \frac{1}{2} \left(\phi_{1-\alpha/2} + \sqrt{2(n-1) - 1} \right)^2 \qquad (2.28)$$

where $\phi_{1-\alpha/2}$ is the quantile of the standard normal law. In the case $n = 100$, at the level 95%, $\alpha = 0.05$, we have:

$$\chi^2_{1-\alpha/2}(n-1) \simeq 129.07, \quad \chi^2_{\alpha/2}(n-1) \simeq 73.77 \qquad (2.29)$$

so that the confidence interval is $\left[0.77S^2, 1.36S^2 \right]$.

2.4.1 The Cochran Theorem

The rigorous justification of the results of this paragraph relies on the following:

Theorem 2.1 (Cochran) *Let* $Y = (Y_1 \ldots Y_n)$ *be an n-dimensional normal random variable,* $Y \sim N(\mathbf{0}, \mathbb{I})$. *Moreover, let* $E_1 \ldots E_s$ *be orthogonal vector subspaces of* $\mathbb{R}^n$, *such that* $\bigoplus_{j=1}^{s} E_j = \mathbb{R}^n$. *We denote* $\Pi_1 \ldots \Pi_s$ *the linear projectors onto such subspaces. Then:*

1. *the random variables* $\Pi_j X$, $j = 1 \ldots s$, *are independent.*
2. *the random variables* $|\Pi_j X|^2$, $j = 1 \ldots s$, *has a chi-square law* $\chi^2(d_j)$, *where* $d_j = dim(E_j)$ *is the dimension of* E_j.

Proof Let $B = \{e_1 \ldots e_n\}$ be the canonical base of $\mathbb{R}^n$, and let's write $Y = \sum_{i=1}^{n} Y_i e_i$.

Moreover, if $\tilde{B}_1 \ldots \tilde{B}_s$ are orthonormal basis of the subspaces $E_1 \ldots E_s$, the set of vectors $\tilde{B} = \tilde{B}_1 \cup \cdots \cup \tilde{B}_s = \{\tilde{e}_1 \ldots \tilde{e}_n\}$ is an orthonormal basis of $\mathbb{R}^n$ and we may write $Y = \sum_{j=1}^{n} \tilde{Y}_j \tilde{e}_j$ where:

$$\begin{pmatrix} \tilde{Y}_1 \\ \cdots \\ \tilde{Y}_n \end{pmatrix} = \begin{pmatrix} (\tilde{e}_1|e_1) & \cdots & (\tilde{e}_1|e_n) \\ \cdots & \cdots & \cdots \\ (\tilde{e}_n|e_1) & \cdots & (\tilde{e}_n|e_n) \end{pmatrix} \begin{pmatrix} Y_1 \\ \cdots \\ Y_n \end{pmatrix}$$

the matrix Γ with matrix elements $\Gamma_{ij} = (\tilde{e}_i|e_j)$ being orthogonal. It follows that the random variable $(\tilde{Y}_1 \ldots \tilde{Y}_n)$ is normal $N(\Gamma \mathbf{0} = \mathbf{0}, \Gamma \mathbb{I} \Gamma^\tau = \mathbb{I})$. Thus, the components $\tilde{Y}_i$ are standard normal and independent.

$$\Pi_j Y = \sum_{i=1}^n (\tilde{e}_i|\Pi_j Y)\tilde{e}_i = \sum_{\tilde{e}_i \in \tilde{B}_j} \tilde{Y}_i \tilde{e}_i$$

implies that the $\Pi_j Y$ are independent. Finally:

$$|\Pi_j Y|^2 = \sum_{\tilde{e}_i \in \tilde{B}_j} \tilde{Y}_i^2 \sim \chi^2(d_j)$$

since the $\tilde{Y}_i$ are standard normal and independent.

2.5 Estimation of a Proportion

Let's assume now that the sample $X = (X_1, \ldots, X_n)$ is made of Bernoulli random variables $B(1, \theta)$ with parameter $\theta \in (0, 1)$. We know that:

$$E_\theta(X_i) = \theta, \quad Var_\theta(X_i) = \theta(1 - \theta)$$

so that the random variable:

$$\mathcal{M} = m(X_1 \ldots X_n) = \frac{\sum_{i=1}^n X_i}{n} \tag{2.30}$$

is an unbiased estimator for θ, that is:

$$E_\theta(\mathcal{M}) = \theta$$

Moreover, for large n, the law of:

$$\frac{\mathcal{M} - \theta}{\sqrt{\theta(1 - \theta)/n}} \tag{2.31}$$

can be approximated by a law $N(0, 1)$, for the central limit theorem. We can thus write:

$$P_\theta \left(-\phi_{1-\frac{\alpha}{2}} \leq \frac{\mathcal{M} - \theta}{\sqrt{\theta(1-\theta)/n}} \leq \phi_{1-\frac{\alpha}{2}} \right) \simeq 1 - \alpha \qquad (2.32)$$

where, as before, $\phi_{1-\frac{\alpha}{2}}$ are the quantiles of the standard normal $N(0, 1)$. We can rewrite the above formula as:

$$P^\theta \left(\mathcal{M} - q_{1-\frac{\alpha}{2}} \frac{\sqrt{\theta(1-\theta)}}{\sqrt{n}} \leq \theta \leq \mathcal{M} + q_{1-\frac{\alpha}{2}} \frac{\sqrt{\theta(1-\theta)}}{\sqrt{n}} \right) \simeq 1 - \alpha$$

In order to build up a confidence interval at the level $1 - \alpha$ for the parameter θ we should solve the inequality:

$$-\phi_{1-\frac{\alpha}{2}} \leq \frac{\mathcal{M} - \theta}{\sqrt{\theta(1-\theta)/n}} \leq \phi_{1-\frac{\alpha}{2}}$$

with respect to θ, which is a simple exercise which we leave to the reader. In general, when n is large enough, the resulting confidence interval can be accurately approximated as:

$$\left[\mathcal{M} - \phi_{1-\frac{\alpha}{2}} \frac{\sqrt{\mathcal{M}(1-\mathcal{M})}}{\sqrt{n}}, \mathcal{M} + \phi_{1-\frac{\alpha}{2}} \frac{\sqrt{\mathcal{M}(1-\mathcal{M})}}{\sqrt{n}} \right] \qquad (2.33)$$

2.6 Cramer-Rao Theorem

We have learnt till now to build up estimators $\mathcal{T}$ for functions $\tau(\boldsymbol{\theta})$, in particular for the mean and the variance, inside a given statistical model. We have seen that some nice properties of an estimator are unbiasedness and consistence. We are going now to explore more deeply the quality of an estimator. Naturally, the *precision* of our estimation will depend on the variance of $\mathcal{T}$, or, in higher dimensions, on its covariance matrix.

We start limiting our attention to real valued estimators and to a one dimensional parameter θ. Later we will generalize to higher dimensions.

We fix some working hypothesis, which are satisfied by a wide class of statistical models, including the exponential ones. First of all, we aussume that the real valued components X_i of the sample $(X_1 \ldots X_n)$ have density $p_\theta(\mathbf{x})$, which we ask to be differentiable with respect to the parameter θ. Moreover, we assume that, for any statistic $\mathcal{T} = t(X_1 \ldots X_n)$, integrable with respect to the density $p_\theta(x_1) \ldots p_\theta(x_n)$, we can exchange integration and differentiation:

$$\frac{\partial}{\partial \theta} \left[E_\theta[\mathcal{T}] \right] = \int dx_1 \ldots dx_n \, t(x_1 \ldots x_n) \frac{\partial}{\partial \theta} \left[p_\theta(x_1) \ldots p_\theta(x_n) \right]$$

A simple manipulation of the above identity leads to the following:

$$\frac{\partial}{\partial \theta}\Big[E_\theta[\mathcal{T}]\Big] = E_\theta\,[\mathcal{T}\,S] \tag{2.34}$$

where we have introduced the **score function**:

$$S = \frac{1}{p_\theta(X_1)\ldots p_\theta(X_n)}\frac{\partial}{\partial \theta}\Big[p_\theta(X_1)\ldots p_\theta(X_n)\Big] = \frac{\partial}{\partial \theta}\log\Big[p_\theta(X_1)\ldots p_\theta(X_n)\Big]$$

which measures the *sensibility* of the density $p_\theta(x)$ with respect to the parameter θ.

The score function statistic has zero mean, as can be proved by using $\mathcal{T} = 1$ in the identity (2.34):

$$\frac{\partial}{\partial \theta}\Big[E_\theta[1]\Big] = 0 = E_\theta\,[\,S] \tag{2.35}$$

The *average sensibility* is thus measured by the variance of the score function, which is called **Fisher Information number**:

$$I(\theta) = E_\theta\left[S^2\right] = E_\theta\left[\left(\frac{\partial}{\partial \theta}\log\Big[p_\theta(X_1)\ldots p_\theta(X_n)\Big]\right)^2\right] \ge 0$$

We will show now that such Fisher information number is related to the *maximum precision* we can expect for one estimator.

We consider now an estimator $\mathcal{T} = t(X_1,\ldots,X_n)$ of a quantity $\tau(\theta)$. If the estimator is **unbiased**, we have:

$$\tau(\theta) = E_\theta[\mathcal{T}] = \int dx_1\ldots dx_n\, t(x_1\ldots x_n)\Big[p_\theta(x_1)\ldots p_\theta(x_n)\Big] \tag{2.36}$$

and, by construction:

$$\frac{d\tau(\theta)}{d\theta} = E_\theta\,[\mathcal{T}\,S] = E_\theta\,[(\mathcal{T} - \tau(\theta))\,S] = Cov\,[\mathcal{T},\,S] \tag{2.37}$$

where we have use the fact that the score function has zero mean.

We can use now Cauchy-Schwartz inequality, which implies the following very interesting result:

$$\left|\frac{d\tau(\theta)}{d\theta}\right|^2 = |Cov\,[\mathcal{T},\,S]\,|^2 \le Var(\mathcal{T})Var(S) \tag{2.38}$$

This is the very important **Cramer-Rao** inequality:

$$Var_\theta(\mathcal{T}) \geq \frac{|\frac{d\tau(\theta)}{d\theta}|^2}{I(\theta)} \qquad (2.39)$$

$I(\theta)$ being the **Fisher Information number**:

$$I(\theta) = E_\theta\left[\left(\frac{\partial}{\partial\theta}\log\left[p_\theta(X_1)\ldots p_\theta(X_n)\right]\right)^2\right]$$

We stress that the Fisher information number is a property of the statistical model, and not of the estimator: nevertheless, it imposes a lower bound to the variance of estimators that can be built up. Naturally, the smallest is the variance, the highest is the precision of the estimation: $I(\theta)$ controls the precision of the estimators.

Definition 2.7 An estimator $\mathcal{T}$ of a quantity $\tau(\theta)$ is called **efficient** if:

$$Var_\theta(\mathcal{T}) = \frac{|\frac{d\tau(\theta)}{d\theta}|^2}{I(\theta)} \qquad (2.40)$$

Let's consider an instructive example: let's assume that the components of the sample X_i are normal with unknown mean θ and **known** variance σ^2. We have:

$$\left(\frac{\partial}{\partial\theta}\log\left[p_\theta(X_1)\ldots p_\theta(X_n)\right]\right)^2 = \left(\sum_{i=1}^{n}\frac{\partial}{\partial\theta}\log\left(\frac{\exp\left(-\frac{(X_i-\theta)^2}{2\sigma^2}\right)}{\sqrt{2\pi}\,\sigma}\right)\right)^2 =$$

$$= \left(\sum_{i=1}^{n}\frac{\partial}{\partial\theta}\left(-\frac{(X_i-\theta)^2}{2\sigma^2}-\log(\sqrt{2\pi}\,\sigma)\right)\right)^2 = \left(\sum_{i=1}^{n}\frac{X_i-\theta}{\sigma^2}\right)^2$$

We get thus, exploiting independence, the following result for the Fisher information:

$$I(\theta) = E_\theta\left[\left(\sum_{i=1}^{n}\frac{X_i-\theta}{\sigma^2}\right)^2\right] = nVar_\theta\left(\frac{X_i-\theta}{\sigma^2}\right) = nE_\theta\left[\left(\frac{X_i-\theta}{\sigma^2}\right)^2\right] = \frac{n}{\sigma^2} \qquad (2.41)$$

independent on θ. On the other hand, if we consider the estimator:

$$\mathcal{M} = m(X_1\ldots X_n) = \frac{\sum_{i=1}^{n}X_i}{n} \qquad (2.42)$$

of $\tau(\theta) = \theta$ (whose derivative is one!), we already know that:

$$Var(\mathcal{M}) = \frac{\sigma^2}{n} = \frac{1}{I(\theta)} \tag{2.43}$$

so that $\mathcal{M}$ is an efficient estimator of the mean: keeping fixed the statistical model, it is not possible to build up an estimator for the mean with variance lower than $\frac{\sigma^2}{n}$.

We present now the general statement of the Cramer-Rao theorem.

Theorem 2.2 (Cramer, Rao) *Let* $\{(\Omega, \mathcal{F}, P_\theta)\}_{\theta \in \Theta}$ *a statistical model and* $X = (X_1 \dots X_n)$ *a sample of rank n such that the following hypothesis hold:*

1. *the law of the components* X_i *of the sample* $(X_1 \dots X_n)$ *has density* $p_\theta(\mathbf{x})$
2. $p_\theta(\mathbf{x})$ *is differentiable with respect to* θ
3. *for any s-dimensional statistic* $\mathcal{T} = t(X_1 \dots X_n)$ *we can write:*

$$\frac{\partial}{\partial\theta}\Big[E_\theta(\mathcal{T}_i)\Big] = \int dx_1 \dots dx_n \, t_i(\mathbf{x}_1 \dots \mathbf{x}_n) \frac{\partial}{\partial\theta}\Big[p_\theta(\mathbf{x}_1) \dots p_\theta(\mathbf{x}_n)\Big]$$

If $\mathcal{T}$ *is an estimator of the quantity* $\tau(\theta)$ *with finite expectation* $E_\theta(\mathcal{T})$, *then the following matrix inequality holds:*

$$Cov_\theta(\mathcal{T}) \geq J(\theta)I^{-1}(\theta)J(\theta)^T \tag{2.44}$$

where $J(\theta)$ *is the Jacobian of* $E_\theta(\mathcal{T})$:

$$J_{ik}(\theta) = \frac{\partial E_\theta(\mathcal{T}_i)}{\partial\theta_k} \tag{2.45}$$

and $I(\theta)$ *is the* **Fisher Information matrix**:

$$I_{ij}(\theta) = E_\theta\left(S_i(X_1 \dots X_n)S_j(X_1 \dots X_n)\right)$$

the **score vector** *being defined by:*

$$S(X_1 \dots X_n) = \frac{\partial}{\partial\theta} \log\Big[p_\theta(X_1) \dots p_\theta(X_n)\Big]$$

Proof Using the constant statistic $\mathcal{T} = 1$, we see that the components of the score vector have zero mean. We introduce the matrix:

$$A_{ik} = E_\theta\left((\mathcal{T}_i - E_\theta(\mathcal{T}_i))\, S_k\right) = E_\theta\left(\mathcal{T}_i\, S_k\right) \tag{2.46}$$

Moreover, by inspection we see that:

$$E_\theta\left(\mathcal{T}_i\, S_k\right) = \frac{\partial E_\theta(\mathcal{T}_i)}{\partial\theta_k} = J_{ik}(\theta) \tag{2.47}$$

This implies that, for each couple of vectors $\mathbf{a} \in \mathbb{R}^s$, $\mathbf{b} \in \mathbb{R}^m$:

$$\mathbf{a} \cdot A\,\mathbf{b} = \mathbf{a} \cdot J(\boldsymbol{\theta})\,\mathbf{b} \qquad (2.48)$$

The left hand side has the explicit form:

$$\mathbf{a} \cdot A\,\mathbf{b} = \sum_{i=1}^{s}\sum_{j=1}^{m} a_i A_{ij} b_j = E_{\boldsymbol{\theta}}\Big[\,(\mathbf{a} \cdot (\mathfrak{I} - E_{\boldsymbol{\theta}}(\mathfrak{I})))\,(S \cdot \mathbf{b}))\,\Big]$$

Cauchy-Schwartz inequality implies:

$$|\mathbf{a} \cdot A\,\mathbf{b}|^2 \leq E_{\boldsymbol{\theta}}\Big(\,(\mathbf{a} \cdot (\mathfrak{I} - E_{\boldsymbol{\theta}}(\mathfrak{I})))^2\,\Big)\,E_{\boldsymbol{\theta}}\Big(\,(S \cdot \mathbf{b}))^2\,\Big)$$

that is:

$$|\mathbf{a} \cdot A\,\mathbf{b}|^2 \leq (\mathbf{a} \cdot \mathrm{Cov}_{\boldsymbol{\theta}}(\mathfrak{I})\,\mathbf{a})\,(\mathbf{b} \cdot I(\boldsymbol{\theta})\,\mathbf{b}) \qquad (2.49)$$

Finally:

$$|\mathbf{a} \cdot J(\boldsymbol{\theta})\,\mathbf{b}|^2 = |\mathbf{a} \cdot A\,\mathbf{b}|^2 \leq (\mathbf{a} \cdot \mathrm{Cov}_{\boldsymbol{\theta}}(\mathfrak{I})\,\mathbf{a})\,(\mathbf{b} \cdot I(\boldsymbol{\theta})\,\mathbf{b})$$

Choosing $\mathbf{b} = I(\boldsymbol{\theta})^{-1} J(\boldsymbol{\theta})^T \mathbf{a}$ and using the symmetry of Fisher information matrix (and of its inverse), we find:

$$(\mathbf{a} \cdot J(\boldsymbol{\theta})\,\mathbf{b})\,\big(\mathbf{a} \cdot J(\boldsymbol{\theta})I(\boldsymbol{\theta})^{-1}J(\boldsymbol{\theta})^T\,\mathbf{a}\big) \leq (\mathbf{a} \cdot \mathrm{cov}_{\boldsymbol{\theta}}(\mathfrak{I})\,\mathbf{a})\,(\mathbf{a} \cdot J(\boldsymbol{\theta}) \cdot \mathbf{b})$$

that is:

$$(\mathbf{a} \cdot \mathrm{cov}_{\boldsymbol{\theta}}(\mathfrak{I})\,\mathbf{a}) \geq \big(\mathbf{a} \cdot J(\boldsymbol{\theta})I(\boldsymbol{\theta})^{-1}J(\boldsymbol{\theta})^T\,\mathbf{a}\big)$$

This completes the proof.

2.7 Maximum Likelihood Estimators (MLE)

We have till now learnt some useful properties of estimators, determining their *precision* in inferring $\tau(\boldsymbol{\theta})$ from a set of data. A natural question is whether there exists a tool to *invent* an estimator for a particular $\tau(\boldsymbol{\theta})$. In the case of mean and variance the actual definition of the estimator is very natural, but there can be situations in which the choice is not so simple. We limit our attention to the case when the quantity $\tau(\boldsymbol{\theta})$ to be estimated is the parameter $\boldsymbol{\theta}$ itself. We assume moreover that the components of the sample $(X_1 \ldots X_n)$ have density $p_{\boldsymbol{\theta}}(\mathbf{x})$.

Given the data $(\mathbf{x}_1 \ldots \mathbf{x}_n)$ let's consider the **likelihood function**:

$$L(\boldsymbol{\theta}; \mathbf{x}_1 \ldots \mathbf{x}_n) \stackrel{def}{=} p_{\boldsymbol{\theta}}(\mathbf{x}_1)\ldots p_{\boldsymbol{\theta}}(\mathbf{x}_n) \qquad (2.50)$$

Intuitively, "$L(\theta; \mathbf{x}_1 \ldots \mathbf{x}_n) d\mathbf{x}_1 \ldots d\mathbf{x}_n$" is the *probability* to obtain precisely the measured data for the value θ of the parameter. We are naturally induced to estimate the unknown parameter as the value θ which **maximizes** such *probability*. This justifies the following:

> **Definition 2.8** We call **maximum likelihood estimator** (MLE) of the parameter θ the statistic:
>
> $$\mathcal{T}_{ML}(X_1 \ldots X_n) = \arg\max_{\theta \in \Theta} L(\theta; X_1 \ldots X_n) = \arg\max_{\theta \in \Theta} p_\theta(X_1) \ldots p_\theta(X_n)$$
> $$(2.51)$$

We observe that such estimator is well defined whenever, for the given data, the function $\theta \to L(\theta; \mathbf{x}_1 \ldots \mathbf{x}_n)$ has a unique maximum.

In order to give a first example, let's consider again the normal sample with unknown mean θ and known variance σ^2. In such case:

$$L(\theta; X_1 \ldots X_n) = \frac{1}{(2\pi\sigma^2)^{n/2}} \exp\left(-\sum_{i=1}^{n} \frac{(X_i - \theta)^2}{2\sigma^2}\right) \qquad (2.52)$$

The maximization of such function with respect to θ leads to the following equation:

$$0 = \frac{\partial}{\partial\theta}\left(-\sum_{i=1}^{n}(X_i - \theta)^2\right) = 2\sum_{i=1}^{n}(X_i - \theta)$$

which implies:

$$\mathcal{T}_{ML}(X_1 \ldots X_n) = \frac{\sum_{i=1}^{n} X_i}{n}$$

which is exactly the empirical mean.

The MLE for some important models are summarized in the following table:

Model	$\mathcal{T}_{ML}$	$E_\theta(\mathcal{T}_{ML})$	$\text{Var}_\theta(\mathcal{T}_{ML})$	$I(\theta)$
Bernoulli	$\frac{\sum_{i=1}^{n} X_i}{n}$	θ	$\frac{\theta(1-\theta)}{n}$	$\frac{n}{\theta(1-\theta)}$
Gaussian	$\begin{pmatrix} \frac{\sum_{i=1}^{n} X_i}{n} \\ \frac{\sum_{i=1}^{n} X_i^2}{n} - \left(\frac{\sum_{i=1}^{n} X_i}{n}\right)^2 \end{pmatrix}$	$\begin{pmatrix} \theta_0 \\ \frac{n-1}{n}\theta_1 \end{pmatrix}$	$\begin{pmatrix} \frac{\theta_1}{n} & 0 \\ 0 & \frac{n-1}{n^2}2\theta_1^2 \end{pmatrix}$	$\begin{pmatrix} \frac{n}{\theta_1} & 0 \\ 0 & \frac{n}{2\theta_1^2} \end{pmatrix}$
Exonential	$\frac{n}{\sum_{i=1}^{n} X_i}$	$\frac{n}{n-1}\theta$	$\frac{n^2}{(n-1)^2(n-2)}\theta^2$	$\frac{n}{\theta^2}$

We observe that the MLE may be not unbiased nor efficient, but they asyntotically have these properties in the limit of large samples.

The following theorems, which we state without proof, provide general results about MLEs. The first result concernes existence of MLEs. For a proof we defer to [2].

Theorem 2.3 (Wald) *If:*

1. *Θ is compact.*
2. *for each $\mathbf{x}$ the density $p_\theta(\mathbf{x})$ is a continuous function of θ.*
3. *$p_\theta(\mathbf{x}) = p_{\theta'}(\mathbf{x})$ if and only if $\theta = \theta'$*
4. *there exists a positive function $K : \mathbb{R}^k \to \mathbb{R}$, such that the random variable $K(X_i)$ has finite expectation and such that, for each $\mathbf{x}$ and θ:*

$$\left| \log \left[\frac{p_\theta(\mathbf{x})}{p_{\theta'}(\mathbf{x})} \right] \right| \leq K(\mathbf{x}) \tag{2.53}$$

Then there exist a maximum likelihood estimator $\mathcal{T}_{ML}(X_1 \ldots X_n)$ that converges almost surely to θ as the rank of the sample increases to $+\infty$.

A stronger result is the following:

Theorem 2.4 (Cramer) *If:*

1. *Θ is open.*
2. *for each $\mathbf{x}$, $p_\theta(\mathbf{x}) \in \mathcal{C}^2(\Theta)$, and it is possible to exchange derivative and expectation.*
3. *$p_\theta(\mathbf{x}) = p_{\theta'}(\mathbf{x})$ if and only if $\theta = \theta'$*
4. *there exists a function $K(\mathbf{x})$ such that $K(X_i)$ has finite expectation and:*

$$\| \nabla_\theta \log \left(p_\theta(\mathbf{x}) \right) \| \leq K(\mathbf{x}) \tag{2.54}$$

for each $\mathbf{x} \in \mathbb{R}^k$.

Then there exists a maximum likelihood estimator $\mathcal{T}_{ML}(X_1 \ldots X_n)$ that converges almost surely to θ as the rank of the sample increases to $+\infty$, and that is asyntotically normal and efficient:

$$\mathcal{T}_{ML}(X_1 \ldots X_n) \xrightarrow[n \to \infty]{a.s.} \theta$$
$$\lim_{n \to \infty} \sqrt{n} \, [\mathcal{T}_{ML}(X_1 \ldots X_n) - \theta] \xrightarrow{\mathcal{L}} Z \sim N \left[0, I(\theta)^{-1} \right] \tag{2.55}$$

2.8 Hypothesis Tests

A typical problem in mathematical statistics is to use the data to confirm or reject an hypothesis relying on a set of data. Once fixed a statistical model, an hypothesis is a statement about the parameter θ. In practice, the statistical hypothesis to be tested, called the **null hypothesis** H_0 (that in general the tester tries to *reject*) can be expressed as:

$$H_0: \quad \boldsymbol{\theta} \in \Theta_0 \tag{2.56}$$

while the **alternative hypothesis** H_1, (that in general the tester tries to *establish*) can be expressed as:

$$H_1: \quad \boldsymbol{\theta} \in \Theta_1 = \Theta - \Theta_0 \tag{2.57}$$

We start from the data $(\mathbf{x}_1 \ldots \mathbf{x}_n)$. Performing a statistical test means choosing a subset $\Omega_R \subset \mathbb{R}^k \times \cdots \times \mathbb{R}^k$, called **critical region**, such that we **reject** the null hypothesis if $(\mathbf{x}_1 \ldots \mathbf{x}_n) \in \Omega_R$:

$$(\mathbf{x}_1 \ldots \mathbf{x}_n) \in \Omega_R \quad \Rightarrow \quad reject \ H_0 \tag{2.58}$$

In such case the conclusion is that H_0 is **not** consistent with the data. Naturally, the randomness in the experiment can lead to errors: if we **reject** H_0 when H_0 is true, we say we do a **type I error**; on the other hand, if we do **not reject** H_0 when H_0 is false, we say that we do a **type II error**.

In most cases, the critical region is expressed in term of a statistic $\mathcal{T} = t(X_1, \ldots, X_n)$, in the form:

$$\Omega_R = \{(\mathbf{x}_1 \ldots \mathbf{x}_n) : t(\mathbf{x}_1 \ldots \mathbf{x}_n) > \mathcal{T}_0\}$$

for a given treshold value $\mathcal{T}_0$.

2.8.1 Student Test

One very common experimental situation is the comparison between the mean of a measured quantity and a reference value, maybe coming from a theoretical study. We assume that the data $(x_1 \ldots x_n)$ can be modeled as realization of a sample $(X_1 \ldots X_n)$, with one dimensional **normal** components. Introducing the statistics $\mathcal{M}$ and $\mathcal{S}^2$, respectively estimators of mean $\mu(\boldsymbol{\theta})$ and variance $\sigma^2(\boldsymbol{\theta})$, we already know that:

$$\mathcal{R} = r(X_1, \ldots, X_n) = \frac{\mathcal{M} - \mu(\boldsymbol{\theta})}{\sqrt{\frac{\mathcal{S}^2}{n}}} \tag{2.59}$$

follows a Student law with $n - 1$ degrees of freedom.

We denote μ_0 the reference value. We test the hypothesis:

$$H_0: \quad \mu(\boldsymbol{\theta}) = \mu_0 \tag{2.60}$$

against:

$$H_1 : \quad \mu(\boldsymbol{\theta}) \neq \mu_0 \qquad (2.61)$$

Naturally we will reject H_0 if the estimated mean is far from μ_0. **If the null hypothesis H_0 is true**, we can calculate:

$$P_\theta \left(\left| \frac{\mathcal{M} - \mu_0}{\sqrt{\frac{S^2}{n}}} \right| > t_{1-\frac{\alpha}{2}}(n-1) \right) = \alpha$$

for any $\alpha \in (0, 1)$. At the **significance level** α, we can define the critical region as:

$$\Omega_R = \{(x_1 \ldots x_n) : |r(x_1 \ldots x_n)| > t_{1-\frac{\alpha}{2}}(n-1)\}$$

Typically chosen values are $\alpha = 0.10, 0.05, 0.01$, corresponding, for large samples, to the quantiles $1.645, 1.96, 2.58$. We note that α is precisely the **probability of type I error**.

Given the data $(x_1 \ldots x_n)$, we can thus immediately calculate the *standardized discrepancy* with respect to the reference value: $r(x_1 \ldots x_n)$. We can also, using the Student law or the normal if the sample is large enough, compute the **p value**:

$$p \ value = P_\theta \left(\left| \frac{\mathcal{M} - \mu_0}{\sqrt{\frac{S^2}{n}}} \right| > r(x_1 \ldots x_n) \right) \qquad (2.62)$$

under the assumption that H_0 is true. If the p value is less than or equal to the significance level α, H_0 is rejected at the significance level α; the p value is the probability to find data *worse* than the ones we have measured if H_0 is true. A small p value means that is very unlikely that H_0 is consistent with the data.

Another important class of hypothesis that are often tested have the form:

$$H_0 : \quad \mu(\boldsymbol{\theta}) \leq \mu_0 \qquad (2.63)$$

Naturally, the alternative is:

$$H_1 : \quad \mu(\boldsymbol{\theta}) > \mu_0 \qquad (2.64)$$

It is clear that we will reject the null hypothesis if we get a mean much bigger than μ_0. In order to be quantitative, we observe that:

$$\mathcal{R} = \frac{\mathcal{M} - \mu_0}{\sqrt{\frac{S^2}{n}}} = \frac{\mathcal{M} - \mu(\boldsymbol{\theta})}{\sqrt{\frac{S^2}{n}}} + \frac{\mu(\boldsymbol{\theta}) - \mu_0}{\sqrt{\frac{S^2}{n}}} \qquad (2.65)$$

is the sum of a Student random variable $t(n-1)$ and a term which is always negative if H_0 is true. Thus:

$$P_\theta \left(\mathcal{R} > t_{1-\alpha}(n-1) \right) \le P_\theta \left(\frac{\mathcal{M} - \mu(\theta)}{\sqrt{\frac{S^2}{n}}} > t_{1-\alpha}(n-1) \right) = \alpha$$

and we may set the critical region:

$$\Omega_R = \{(x_1 \ldots x_n) : r(x_1 \ldots x_n) > t_{1-\alpha}(n-1)\}$$

defines a statistical test of the hypothesis (2.63) whose probability of I type error is α.

In several situations two different estimations of averages of **independent normal** samples $X = (X_1 \ldots X_n)$ and $Y = (Y_1 \ldots Y_m)$ are compared. We will limit our attention to the situation in which the two independent samples share the same value for the variance.

In the simplest case, the hypothesis that the two means are equal:

$$H_0 : \quad \mu_X(\theta) = \mu_Y(\theta) \tag{2.66}$$

is tested against the alternative:

$$H_1 : \quad \mu_X(\theta) \ne \mu_Y(\theta) \tag{2.67}$$

Using Cochran theorem, it is simple to show that, if H_0 is true, the random variable:

$$\mathcal{T} = \frac{\mathcal{M}_X - \mathcal{M}_Y}{\sqrt{\frac{1}{n} + \frac{1}{m}} \sqrt{\frac{(n-1)S_X^2 + (m-1)S_Y^2}{n+m-2}}}$$

follows a Student law with $n + m - 2$ degrees of freedom $t(n + m - 2)$. The above notation is precisely the same we have used throughout this chapter a part from a label to distinguish the two samples: $\mathcal{M}_X = \frac{1}{n} \sum_{i=1}^{n} X_i$ and so on. We have thus:

$$P_\theta \left(|\mathcal{T}| > t_{1-\frac{\alpha}{2}}(n + m - 2) \right) = \alpha \tag{2.68}$$

providing a critical region at the significance level α of the form:

$$\Omega_R = \{(x_1 \ldots x_n; y_1 \ldots y_m) : |t| > t_{1-\frac{\alpha}{2}}(n + m - 2)\}$$

where $t = \mathcal{T}(x_1 \ldots x_n; y_1 \ldots y_m)$. Intuitively, if the two estimations of the means turn out to be "too different", we reject the hypothesis.

If we have to test the hypothesis:

$$H_0: \quad \mu_X(\boldsymbol{\theta}) \le \mu_Y(\boldsymbol{\theta}) \tag{2.69}$$

we rely on the observation that:

$$
\mathcal{T} = \frac{\mathcal{M}_X - \mathcal{M}_Y}{\sqrt{\frac{1}{n} + \frac{1}{m}} \sqrt{\frac{(n-1)S_X^2 + (m-1)S_Y^2}{n+m-2}}} =
$$

$$
= \frac{(\mathcal{M}_X - \theta_{0X}) - (\mathcal{M}_Y - \theta_{0Y})}{\sqrt{\frac{1}{n} + \frac{1}{m}} \sqrt{\frac{(n-1)S_X^2 + (m-1)S_Y^2}{n+m-2}}} + \frac{\theta_{0X} - \theta_{0Y}}{\sqrt{\frac{1}{n} + \frac{1}{m}} \sqrt{\frac{(n-1)S_X^2 + (m-1)S_Y^2}{n+m-2}}}
$$

is the sum of a random variable with law $t(n + m - 2)$ and a term which, if H_0 is true, is always negative or equal to zero. Therefore:

$$P_\theta \left(\mathcal{T} > t_{1-\alpha}(n+m-2) \right) = \alpha \tag{2.70}$$

and the crital region for a test at the significance level α is:

$$\Omega_R = \{ (x_1 \dots x_n; y_1 \dots y_m) : t > t_{1-\alpha}(n+m-2) \}$$

where $t = \mathcal{T}(x_1 \dots x_n; y_1 \dots y_m)$.

2.8.2 Chi-Squared Test

The **Chi-Squared test**, or **Goodness-of-Fit test**, due to Pearson, is a test of the hypothesis:

$$H_0: \quad \boldsymbol{\theta} = \boldsymbol{\theta}_0$$

aiming to verify whether the probability density $p_{\theta_0}(\mathbf{x})$, specified by the value $\boldsymbol{\theta}_0$ of the parameter, is a good description of the experiment we have made, given a set of data $(\mathbf{x}_1 \dots \mathbf{x}_n)$. The starting point is a partition of the range of the measurements, $\mathbb{R}^k$, in a **finite** family $\{E_j\}_{j=1}^r$ of **outcomes**, mutually disjoint such that $\bigcup_{j=1}^r E_j = \mathbb{R}^k$. The basic idea of the test is to compare the **theoretical frequencies**:

$$p_j(\boldsymbol{\theta}_0) = P_{\theta_0}(X_i \in E_j) = \int_{E_j} d\mathbf{x}\, p_{\theta_0}(\mathbf{x})$$

with the **empirical frequencies**:

$$f_j = \frac{\sum_{i=1}^n 1_{E_j}(\mathbf{x}_i)}{n}$$

giving the number of measurements fallen in the set E_j.

As usual, we interpret the numbers f_j as realizations of the statistics:

$$\mathcal{N}_j = n_j(X_1 \ldots X_n) = \frac{\sum_{i=1}^n 1_{E_j}(X_i)}{n}$$

The discrepancy between theoretical and empirical frequencies builds up the **Pearson random variable**:

$$\mathcal{P} = p(X_1, \ldots, X_n) = \sum_{j=1}^r n \frac{\left(\mathcal{N}_j - p_j(\boldsymbol{\theta}_0)\right)^2}{p_j(\boldsymbol{\theta}_0)} \tag{2.71}$$

The key result is the following, which we will prove in the problems section:

Theorem 2.5 *If the hypothesis:*

$$H_0 : \quad \boldsymbol{\theta} = \boldsymbol{\theta}_0$$

is true, the Pearson random variable converges in distribution to a random variable $\chi^2(r-1)$, as the rank of the sample tends to $+\infty$.

Thus, assuming H_0 true, if the sample is large enough, we have:

$$P_{\boldsymbol{\theta}_0}\left(\mathcal{P} \geq \chi_{1-\alpha}^2(r-1)\right) = \alpha \tag{2.72}$$

so that we can define the critical region for the test of H_0 at the significance level α.

$$\Omega_R = \left\{(\mathbf{x}_1 \ldots \mathbf{x}_n) : p(\mathbf{x}_1 \ldots \mathbf{x}_n) > \chi_{1-\alpha}^2(r-1)\right\} \tag{2.73}$$

2.8.3 Kolmogorov-Smirnov Test

The weak point of the Pearson χ^2 test is the necessity of introducing the partition $\{E_j\}_{j=1}^r$ of the outcomes, which is quite arbitrary. In this section we will describe a different approach due to Kolmogorov and Smirnov. The aim is again to test the hypothesis:

$$H_0 : \quad \boldsymbol{\theta} = \boldsymbol{\theta}_0$$

We will assume to deal with a sample $(X_1 \ldots X_n)$ made of one-dimensional random variables with cumulative distribution function $F_{\boldsymbol{\theta}} : \mathbb{R} \to [0, 1]$ that is **continuous** and **strictly increasing** for all $\boldsymbol{\theta}$.

We introduce now the **empirical cumulative distribution function** of the sample:

$$\tilde{F}_n(x) = \frac{1}{n} \sum_{i=1}^{n} \Theta(x - X_i)$$

where Θ is the Heaviside distribution, $\Theta(x) = 1$ if $x > 0$ and $\Theta(x) = 0$ if $x < 0$. $\tilde{F}_n(x)$, for all $x \in \mathbb{R}$, is a random variable counting the number of outcomes smaller or equal to x. The following calculation shows that $\tilde{F}_n(x)$ is an unbiased and consistent estimator of the "true" cumulative distribution function F_θ:

$$E_\theta[\tilde{F}_n(x)] = \frac{1}{n} \sum_{i=1}^{n} E_\theta[\Theta(x - X_i)] = P_\theta(X \le x) = F_\theta(x)$$

$$\mathrm{var}_\theta[\tilde{F}_n(x)] = \frac{1}{n^2} \sum_{ij=1}^{n} E_\theta[\Theta(x - X_i)\Theta(x - X_j)] - F_\theta(x)^2 = \frac{F_\theta(x)(1 - F_\theta(x))}{n}$$

Moreover, we are going now to show the following important result:

Theorem 2.6 (Glivenko-Cantelli) *The **Kolmogorov-Smirnov random variable**:*

$$\sup_{x \in \mathbb{R}} |\tilde{F}_n(x) - F_\theta(x)|$$

converges in probability to zero, that is:

$$\lim_{n \to \infty} P_\theta\left(\sup_{x \in \mathbb{R}} |\tilde{F}_n(x) - F_\theta(x)| \le \epsilon \right) = 1$$

$\forall \epsilon > 0.$

Proof Let's fix $\epsilon > 0$, choose $k \in \mathbb{N}$, $k \ge \frac{1}{2\epsilon}$ and consider the points $x_j = F_\theta^{-1}\left(\frac{j}{k}\right)$ with $j = 0 \ldots k$. Then:

$$F_\theta(x_{j+1}) - F_\theta(x_j) = \frac{j+1}{k} - \frac{j}{k} = \frac{1}{k} \le \epsilon$$

As we have observed above, in each point x_j the empirical cumulative distribution function $\tilde{F}_n(x_j)$ converges in probability to $F_\theta(x_j)$. Then the random variable:

$$\Delta_k = \max_{j=0\ldots k} |\tilde{F}_n(x_j) - F_\theta(x_j)|$$

converges in probability to zero. Since $\forall x \in \mathbb{R}$ there exists one and only one j such that $x \in [x_{j-1}, x_j)$ we can write:

$$\tilde{F}_n(x) - F_\theta(x) \le \tilde{F}_n(x_j) - F_\theta(x_{j-1}) = \tilde{F}_n(x_j) - F_\theta(x_j) + F_\theta(x_j) - F_\theta(x_{j-1})$$

$$|\tilde{F}_n(x) - F_\theta(x)| \le |\tilde{F}_n(x_j) - F_\theta(x_j)| + |F_\theta(x_j) - F_\theta(x_{j-1})| \le \Delta_k + \epsilon$$

The fact that the last member is independent of x allows to write:

$$\sup_{\mathbb{R}} |\tilde{F}_n(x) - F_\theta(x)| \le \Delta_k + \epsilon$$

which implies:

$$P_\theta \left(\sup_{x \in \mathbb{R}} |\tilde{F}_n(x) - F_\theta(x)| \le 2\epsilon \right) \ge P_\theta \left(\Delta_k + \epsilon \le 2\epsilon \right) = P_\theta \left(\Delta_k \le \epsilon \right)$$

This completes the proof since Δ_k converger in probability to zero:

$$\lim_{n \to \infty} P_\theta \left(\sup_{\mathbb{R}} |\tilde{F}_n(x) - F_\theta(x)| \le 2\epsilon \right) \ge 1$$

Let's consider now the random variable:

$$\sqrt{n} \sup_{x \in \mathbb{R}} |\tilde{F}_n(x) - F_\theta(x)| = \sqrt{n} \sup_{x \in \mathbb{R}} |\frac{1}{n} \sum_{i=1}^{n} \Theta(x - X_i) - F_\theta(x)|$$

for the given sample. Since F_θ is invertible by construction, we can write:

$$\sqrt{n} \sup_{\mathbb{R}} |\tilde{F}_n(x) - F_\theta(x)| = \sqrt{n} \sup_{[0,1]} \left| \tilde{F}_n(F_\theta^{-1}(t)) - t \right| = \sup_{[0,1]} \left| \frac{1}{\sqrt{n}} \sum_{i=1}^{n} \Theta(t - F_\theta(X_i)) - \sqrt{n}t \right|$$

Now, let's define:

$$\tilde{B}_t^{(n)} = \frac{1}{\sqrt{n}} \sum_{i=1}^{n} \Theta(t - F_\theta(X_i)) - \sqrt{n}t$$

The key point is that $F_\theta(X_1), \ldots, F_\theta(X_n)$ are independent and **uniform** in $(0, 1)$, as we have shown in the first chapter. It is immediate to see that:

$$\tilde{B}_0^{(n)} = \tilde{B}_1^{(n)} = 0$$

Moreover:

$$E\left[\tilde{B}_t^{(n)} \right] = \frac{1}{\sqrt{n}} \sum_{i=1}^{n} \int_0^1 du \Theta(t - u) - \sqrt{n}t = 0$$

and, if $s < t$:

$$E\left[\tilde{B}_t^{(n)}\tilde{B}_s^{(n)}\right] = \frac{1}{n}\sum_i \int_0^1 du\,\Theta(t-u)\Theta(s-v)du+$$

$$+\frac{1}{n}\sum_{i\neq j=1}^n \int_0^1 du \int_0^1 dv\,\Theta(t-u)\Theta(s-v) + nts+$$

$$-s\sum_{i=1}^n \int_0^1 du\,\Theta(t-u) - t\sum_{i=1}^n \int_0^1 dv\,\Theta(s-v) =$$

$$= s + (n-1)ts + nts - nts - nts =$$

$$= s(1-t)$$

Finally, the central limit theorem guarantees that $\tilde{B}_t$, in the limit $n \to +\infty$ becomes normal. When we will introduce the theory of stochastic processes, we will call the process:

$$\tilde{B}_t = \lim_{n\to+\infty} \tilde{B}_t^{(n)}$$

brownian bridge. We have thus shown that:

$$\lim_{n\to\infty} \sqrt{n} \sup_{\mathbb{R}} |\tilde{F}_n(x) - F_\theta(x)| = \sup_{[0,1]} |\tilde{B}(t)| \tag{2.74}$$

where $\tilde{B}(t)$ is a brownian bridge. This is very useful since the following technical result, of which we will omit the proof, holds:

$$P\left(\sup_{[0,1]} |\tilde{B}(t)| \leq x\right) = 1 - 2\sum_{k=1}^{\infty}(-1)^{k-1}e^{-2k^2x^2} \tag{2.75}$$

The reader may refer to the following table of the quantiles of the random variable $\sup_{[0,1]} |\tilde{B}(t)|$ (Table 2.1):

Table 2.1 Quantiles $D_{1-\alpha}$ of the random variable $\sup_{[0,1]} |\tilde{B}(t)|$

$1-\alpha$	$D_{1-\alpha}$
0.99	1.627
0.98	1.518
0.95	1.358
0.90	1.222
0.85	1.138
0.80	1.073

The critical region of the Kolmogorov-Smirnov test is:

$$\Omega_R = \left\{ (x_1 \ldots x_n) \ : \ \sqrt{n} \sup_{\mathbb{R}} |\tilde{F}_n(x) - F_{\theta_0}(x)| > D_{1-\alpha} \right\} \qquad (2.76)$$

Given the data, the tester, assuming that $\theta = \theta_0$ evaluates the empirical cumulative distribution function $\tilde{F}_n(x)$ and finds the real number $\sqrt{n} \sup_{\mathbb{R}} |\tilde{F}_n(x) - F_{\theta_0}(x)|$; if such positive number is bigger than $D_{1-\alpha}$, the tester rejects the hypothesis at the significance level α.

2.9 Estimators of Covariance and Correlation

During experiments a very important issue is the existence of correlations among different quantities that are measured. Let's consider the the simplest situation, when only two quantities are measured: this results into two sets of data, $(x_1 \ldots x_n)$ and $(y_1 \ldots y_n)$, which we view as realizations of two samples $(X_1 \ldots X_n)$ e $(Y_1 \ldots Y_n)$.

We wish to estimate the covariance:

$$\mathrm{Cov}_\theta(X_i Y_i) = E_\theta(X_i Y_i) - E_\theta(X_i) E_\theta(Y_i) = \mu_{XY}(\theta) - \mu_X(\theta)\mu_Y(\theta) \qquad (2.77)$$

Let's define the estimator:

$$\mathcal{C} = \frac{n}{n-1} \mathcal{M}_{XY}(X_1 Y_1 \ldots X_n Y_n) - \frac{n}{n-1} \mathcal{M}_X(X_1 \ldots X_n)\mathcal{M}_Y(Y_1 \ldots Y_n) =$$
$$= \frac{1}{n-1} \sum_{i=1}^n X_i Y_i - \frac{1}{n(n-1)} \left(\sum_{i=1}^n X_i \right) \left(\sum_{i=1}^n Y_i \right) \qquad (2.78)$$

This is an unbiased estimator for $\mu_{XY}(\theta) - \mu_X(\theta)\mu_Y(\theta)$, as can be seen from the following calculation:

$$E_\theta(\mathcal{C}) = \frac{\sum_{i=1}^n E_\theta(X_i Y_i)}{n-1} - \frac{\sum_{ij=1}^n E_\theta(X_i Y_j)}{n(n-1)} =$$
$$= E_\theta(X_i Y_i) - E_\theta(X_i) E_\theta(Y_i) = \mu_{XY}(\theta) - \mu_X(\theta)\mu_Y(\theta)$$

Moreover, the law of large numbers guarantees that $\mathcal{C}$ is also consistent.

A very interesting quantitative information about correlation, very often used in data analysis, is the **Pearson correlation coefficient**:

$$\rho(\theta) = \frac{\mu_{XY}(\theta) - \mu_X(\theta)\mu_Y(\theta)}{\sqrt{\sigma_X^2(\theta)\,\sigma_Y^2(\theta)}}$$

which is a real number, $-1 \le \rho(\boldsymbol{\theta}) \le 1$, is zero if the quantities are non correlated and reaches the value ± 1 when there exists a linear relationship between the two quantities.

A typical estimator for $\rho(\boldsymbol{\theta})$ is:

$$\mathcal{R} = \frac{\mathcal{M}_{XY} - \mathcal{M}_X \mathcal{M}_Y}{\sqrt{\mathcal{S}_X^2 \, \mathcal{S}_Y^2}} \qquad (2.79)$$

This natural estimator is a quite complicated function of the samples: it is highly non trivial to evaluate its expectation or to build up confidence intervals. It is useful to introduce here a well established technique, the propagation of errors, which will help us to study the estimator $\mathcal{R}$. The first observation is that:

$$\mathcal{R} = g\left(\mathcal{M}_X, \mathcal{M}_Y, \mathcal{M}_{X^2}, \mathcal{M}_{Y^2}, \mathcal{M}_{XY}\right)$$

where:

$$g(x_1, x_2, x_3, x_4, x_5) = \frac{x_5 - x_1 x_2}{\sqrt{(x_3 - x_1^2)(x_4 - x_2^2)}}$$

and we know the properties of the statistics $\mathcal{M}_X, \mathcal{M}_Y, \mathcal{M}_{X^2}, \mathcal{M}_{Y^2}, \mathcal{M}_{XY}$. What can we learn about $\mathcal{R}$?

The approach we will follow relies on the important theorem:

Theorem 2.7 (Propagation of errors) *Let $\{Z_n\}_{n=0}^{\infty}$ be a sequence of k dimensional random variables converging almost surely to a constant vector $z \in \mathbb{R}^k$ and such as the sequence:*

$$\frac{Z_n - z}{1/\sqrt{n}} \qquad (2.80)$$

converges in distribution to a normal random variable $N(0, \Sigma)$ for a given matrix Σ. If $g : \mathbb{R}^k \to \mathbb{R}$ is a function of class C^1 in a neighborhood of z, then the sequence:

$$\frac{g(Z_n) - g(z)}{1/\sqrt{n}} \qquad (2.81)$$

converges in distribution to a normal random variable $N(0, \nabla g(z) \Sigma \nabla g(z)^T)$.

Proof The proof relies on a first order Taylor expansion with Lagrange rest:

$$g(Z_n(\omega)) = g(z) + \nabla g(Z_n^*(\omega))(Z_n(\omega) - z)$$

where $Z_n^*(\omega)$ lies, for all ω, between z and $Z_n(\omega)$. Exploiting the continuity of ∇g (g is of class C^1 by construction), we have thus:

$$\lim_{n \to \infty} \sqrt{n}\,(g(Z_n) - g(z)) = \nabla g(z) \lim_{n \to \infty} \sqrt{n}\,(Z_n - z)$$

and the thesis follows from the fact that the right hand side has law $N(0, \nabla g(z) \Sigma \nabla g(z)^T)$.

Now, the sequence $(\mathcal{M}_X, \mathcal{M}_Y, \mathcal{M}_{X^2}, \mathcal{M}_{Y^2}, \mathcal{M}_{XY})$ converges almost surely to the limit $(\mu_X(\theta), \mu_Y(\theta), \mu_{X^2}(\theta), \mu_{Y^2}(\theta), \mu_{XY}(\theta))$ when the rank of the samples tends to $+\infty$. Since:

$$\mathcal{R} = g(\mathcal{M}_X, \mathcal{M}_Y, \mathcal{M}_{X^2}, \mathcal{M}_{Y^2}, \mathcal{M}_{XY})$$

where:

$$g(x_1, x_2, x_3, x_4, x_5) = \frac{x_5 - x_1 x_2}{\sqrt{(x_3 - x_1^2)(x_4 - x_2^2)}}$$

is continuous $(\mu_X(\theta), \mu_Y(\theta), \mu_{X^2}(\theta), \mu_{Y^2}(\theta), \mu_{XY}(\theta))$, then $\mathcal{R}$ converges almost surely to $\rho(\theta)$ (the interested reader can try to show this intuitive continuous mapping theorem). This guarantees the consistency of the estimator, since almost sure convergence implies convergence in probability. Moreover, since:

$$\lim_{n \to \infty} E_\theta(\mathcal{R}) = \rho(\theta)$$

$\mathcal{R}$ is also an asyntotically unbiased estimator.

If the components (X_i, Y_i) follow a normal law, we can also use the propagation of errors to find the law of $\mathcal{R}$ and to provide confidence intervals for $\rho(\theta)$. In order to simplify the notations, we let $\mu_X(\theta) = 0$ and $\mu_Y(\theta) = 0$.

For the central limit theorem, the random variable:

$$\lim_{n \to \infty} \sqrt{n} \left[\begin{pmatrix} \mathcal{M}_X \\ \mathcal{M}_Y \\ \mathcal{M}_{X^2} \\ \mathcal{M}_{Y^2} \\ \mathcal{M}_{XY} \end{pmatrix} - \begin{pmatrix} 0 \\ 0 \\ \sigma_X^2(\theta) \\ \sigma_Y^2(\theta) \\ \rho(\theta)\sigma_X(\theta)\sigma_Y(\theta) \end{pmatrix} \right]$$

follows a normal law with covariance matrix Σ whose explicit form is:

$$\Sigma = \begin{pmatrix} \sigma_X^2 & 2\rho\sigma_X\sigma_Y & 0 & 0 & 0 \\ \rho\sigma_X\sigma_Y & \sigma_Y^2 & 0 & 0 & 0 \\ 0 & 0 & 2\sigma_X^4 & 2\rho^2\sigma_X^2\sigma_Y^2 & 2\rho\sigma_X^3\sigma_Y \\ 0 & 0 & 2\rho^2\sigma_X^2\sigma_Y^2 & 2\sigma_Y^4 & 2\rho\sigma_Y^3\sigma_X \\ 0 & 0 & 2\rho\sigma_X^3\sigma_Y & 2\rho\sigma_Y^3\sigma_X & (1+\rho^2)\sigma_X^2\sigma_Y^2 \end{pmatrix}$$

Moreover:

$$\nabla g(0, 0, \sigma_X^2(\boldsymbol{\theta}), \sigma_Y^2(\boldsymbol{\theta}), \rho(\boldsymbol{\theta})\sigma_X(\boldsymbol{\theta})\sigma_Y(\boldsymbol{\theta})) = \begin{pmatrix} 0 \\ 0 \\ -\frac{\rho(\boldsymbol{\theta})}{\sigma_X^2(\boldsymbol{\theta})} \\ -\frac{\rho(\boldsymbol{\theta})}{\sigma_Y^2(\boldsymbol{\theta})} \\ \frac{1}{\sigma_X(\boldsymbol{\theta})\sigma_Y(\boldsymbol{\theta})} \end{pmatrix}$$

Since $\nabla g \Sigma \nabla g = (1 - \rho(\boldsymbol{\theta})^2)^2$, the propagation of errors theorem guarantees that:

$$\lim_{n \to \infty} \sqrt{n}\,(\mathcal{R} - \rho(\boldsymbol{\theta})) \sim N\left(0, \left(1 - \rho(\boldsymbol{\theta})^2\right)^2\right) \qquad (2.82)$$

The reader can verify that the same result can be obtained also when $\mu_X(\boldsymbol{\theta})$ and $\mu_Y(\boldsymbol{\theta})$ do not vanish.

We have thus:

$$P_\theta\left(-\phi_{1-\frac{\alpha}{2}} \le \sqrt{n}\,\frac{\mathcal{R} - \rho(\boldsymbol{\theta})}{1 - \rho(\boldsymbol{\theta})^2} \le \phi_{1-\frac{\alpha}{2}}\right) \simeq 1 - \alpha \qquad (2.83)$$

and a confidence interval at the level $1 - \alpha$ for $\rho(\boldsymbol{\theta})$ turns out to be:

$$\left[\frac{1 - \sqrt{1 - 4z\mathcal{R} + 4z^2}}{2z}, \frac{\sqrt{1 + 4z\mathcal{R} + 4z^2} - 1}{2z}\right] \qquad (2.84)$$

where $z = \frac{\phi_{1-\frac{\alpha}{2}}}{\sqrt{n}}$.

2.10 Linear Regression

We conclude this chapter with a brief review of the well known linear regression, which is widely used in applied science, data analysis and machine learning. It is very common, in several applications, to guess an affine-linear relation between two quantities, say X and Y. The quantity X is usually called the **input variable**, and can be controlled by the experimentalist, who chooses n-values, $(x_1, \ldots, x_n)$ and, correspondingly, performes n measurements of the **response variable**, Y. The response variable is random and the experimentalist will obtain n data $(y_1, \ldots, y_n)$. If we expect a linear-affine relation between X and Y, the simplest way to describe the experiment using the language of probability theory is to model $(y_1, \ldots, y_n)$ as realizations of n random variables of the form:

$$Y_i = a + bx_i + \sigma \varepsilon_i \qquad (2.85)$$

where the x_i enter simply as real parameters, while $\varepsilon_i \sim N(0, 1)$ are standard normal independent random variables; the coefficients a and b of the linear-affine relation, and the measurement error σ, are to be inferred from the data.

We are going now to show how to build up estimations of a, b, and σ, using as starting point the input parameters $(x_1, \ldots, x_n)$ and the data $(y_1, \ldots, y_n)$.

If we organize the data in couples $(x_1, y_1) \ldots (x_n, y_n)$, the most natural strategy is to find the values of a and b minimizing the quantity:

$$F(a, b) = \sum_{i=1}^{n} |y_i - a - bx_i|^2 \tag{2.86}$$

We can write the above function in a more geometrical way as follows:

$$F(a, b) = \left| \mathbf{y} - M \begin{pmatrix} a \\ b \end{pmatrix} \right|^2 \tag{2.87}$$

where $\mathbf{y} = (y_1, \ldots, y_n)$ and $M \in M_{n \times 2}(\mathbb{R})$ is the matrix:

$$\begin{pmatrix} 1 & x_1 \\ \cdots & \cdots \\ 1 & x_n \end{pmatrix} \tag{2.88}$$

whose columns are linearly independent provided that $(x_1 \ldots x_n)$ are not all equals. As (a, b) vary, the set of points $M \begin{pmatrix} a \\ b \end{pmatrix}$ is the plane E_1 in $\mathbb{R}^n$ spanned by the columns of M. Thus:

$$\min_{(a,b) \in \mathbb{R}^2} F(a, b) = \min_{\mathbf{p} \in E_1} |\mathbf{y} - \mathbf{p}|^2 \tag{2.89}$$

so that elementary geometry implies that the minimum is reached when $\mathbf{p} = \Pi_1 \mathbf{y}$, Π_1 being the projector onto the plane E_1, whose explicit form is the following:

$$\Pi_1 = M (M^T M)^{-1} M^T \tag{2.90}$$

We have thus:

$$\mathbf{p} = \Pi_1 \mathbf{y} = M (M^T M)^{-1} M^T \mathbf{y} \tag{2.91}$$

which, keeping in mind that $\mathbf{p} = M \begin{pmatrix} a \\ b \end{pmatrix}$, leads to the estimator:

$$\begin{pmatrix} \mathcal{A} \\ \mathcal{B} \end{pmatrix} (Y_1 \ldots Y_n) = (M^T M)^{-1} M^T Y \tag{2.92}$$

or, more explicitly:

$$A(Y_1 \ldots Y_n) = \frac{\mathcal{M}_{X^2}\mathcal{M}_Y - \mathcal{M}_X\mathcal{M}_{XY}}{\mathcal{M}_{X^2} - \mathcal{M}_X^2}$$

$$B(Y_1 \ldots Y_n) = \frac{\mathcal{M}_{XY} - \mathcal{M}_X\mathcal{M}_Y}{\mathcal{M}_{X^2} - \mathcal{M}_X^2}$$

(2.93)

We stress that the quantities $\mathcal{M}_X$ and $\mathcal{M}_{X^2}$ are **not** random, depending only on the input data. The random variables (2.93) are unbiased estimators of the parameters (a, b); in fact:

$$E\left[\begin{pmatrix} A \\ B \end{pmatrix}\right] = (M^T M)^{-1} M^T E[Y] = (M^T M)^{-1} M^T M \begin{pmatrix} a \\ b \end{pmatrix} = \begin{pmatrix} a \\ b \end{pmatrix}$$

(2.94)

We have still to build up an estimator for σ^2. The idea is that such parameter determines the discrepancy between the data $(y_1, \ldots, y_n)$ and the points of the **regression line** $(a + bx_1, \ldots, a + bx_n)$. It is thus natural to interrelate such parameter to the minimum of the function:

$$\min_{\mathbf{p} \in E_1} |\mathbf{y} - \mathbf{p}|^2 = |\Pi_2 \mathbf{y}|^2$$

(2.95)

Π_2 being the projector onto the $n - 2$-dimensional orthogonal complement of the plane E_1. Such quantity can be interpreted as a realization of the random variable:

$$S^2(Y_1 \ldots Y_n) = \frac{|\Pi_2 Y|^2}{n - 2} = \frac{\sum_i |Y_i - A - B X_i|^2}{n - 2}$$

(2.96)

which is an unbiased estimator for σ^2 since:

$$E\left[|\Pi_2 Y|^2\right] = \sigma^2 E\left[|\Pi_2 \epsilon|^2\right] = \sigma^2 (n - 2)$$

(2.97)

We can also estimate confidence intervals for the parameters a, b, σ^2 relying on the following result:

Theorem 2.8 *The random variables A and B are independent from S^2. Moreover:*

$$\frac{S^2}{\sigma^2} \sim \frac{\chi^2(n-2)}{n-2}$$

$$\frac{A - a}{\sqrt{m_a}\, S} \sim t(n-2)$$

$$\frac{B - b}{\sqrt{m_b}\, S} \sim t(n-2)$$

(2.98)

where:

$$m_a = \frac{\mathcal{M}_{X^2}}{n(\mathcal{M}_{X^2} - \mathcal{M}_X^2)}, \quad m_b = \frac{1}{n(\mathcal{M}_{X^2} - \mathcal{M}_X^2)}$$

(2.99)

Proof Since the random variable ϵ and the linear projectors Π_1, Π_2 satisfy the hyphothesis of Cochran theorem, the random variables $\Pi_1\epsilon$ and $\Pi_2\epsilon$ are independent. Moreover, Cochran theorem guarantees that:

$$|\Pi_2\epsilon|^2 \sim \chi^2(n-2) \tag{2.100}$$

since the subspace onto which Π_2 projects has dimension $n-2$. Thus $\frac{\mathcal{S}^2}{\sigma^2}\frac{|\Pi_2\epsilon|^2}{n-2} \sim$ $\frac{\chi^2(n-2)}{n-2}$. Since the covariance matrix of the random variable $(M^TM)^{-1}M^T\mathbf{Y}$ is:

$$= Cov[(M^TM)^{-1}M^TY] = (M^TM)^{-1}M^TCov[Y]M(M^TM)^{-1} = \sigma^2(M^TM)^{-1} \tag{2.101}$$

we conclude that $\mathcal{A} \sim N(a, m_a\sigma^2), \mathcal{B} \sim N(b, m_b\sigma^2)$ where $m_a = \left[(M^TM)^{-1}\right]_{11} = \frac{\mathcal{M}_{x^2}}{n(\mathcal{M}_{x^2}-\mathcal{M}_X^2)}$ and $m_b = \left[(M^TM)^{-1}\right]_{22} = \frac{1}{n(\mathcal{M}_{x^2}-\mathcal{M}_X^2)}$. The definition of the Student law together with the independence of $\mathcal{A}$ and $\mathcal{B}$ of $\mathcal{S}$ completes the proof.

We are now able to provide confidence intervals of level $1-\alpha$ for σ^2:

$$\left[\frac{n-2}{\chi^2_{1-\frac{\alpha}{2}}(n-2)}\mathcal{S}^2, \frac{n-2}{\chi^2_{\frac{\alpha}{2}}(n-2)}\mathcal{S}^2\right] \tag{2.102}$$

and for a and b:

$$\left[\mathcal{A} - \sqrt{m_a}\,\mathcal{S}\,t_{1-\frac{\alpha}{2}}(n-2), \mathcal{A} + \sqrt{m_a}\,\mathcal{S}\,t_{1-\frac{\alpha}{2}}(n-2)\right]$$
$$\left[\mathcal{B} - \sqrt{m_b}\,\mathcal{S}\,t_{1-\frac{\alpha}{2}}(n-2), \mathcal{B} + \sqrt{m_b}\,\mathcal{S}\,t_{1-\frac{\alpha}{2}}(n-2)\right] \tag{2.103}$$

2.11 Further Readings

This chapter contains all the notions necessary to perform standard statistical data analysis. Further topics in Statistics are covered in many excellent textbooks, like, for example [1–3].

Problems

2.1 An estimator
Consider a sample $(X_1 \ldots X_n)$ with iid components following the law $p_\theta(x) = \frac{1}{\theta}\chi_{[0,\theta]}(x)$. The statistics $\mathcal{T}(X_1 \ldots X_n) = \max(X_1 \ldots X_n)$ can be used to estimate θ. Is it an unbiased estimator? Is it consistent?

2.2 Mean squared error

Given an integrable and square-integrable estimator $\mathcal{T}$ of a quantity $\tau(\theta)$. Provide an expression for the **Mean Squared Error** of the estimator:

$$MSE(\mathcal{T}) = E\left[(\mathcal{T} - \tau(\theta))^2\right] \tag{2.104}$$

in terms of the variance of the estimator itself. What is the role of bias?

2.3 Cochran theorem

Use Cochran theorem to show that, if the sample is made of normal random variables $X_i \sim N(\theta_0, \theta_1)$, then, given the estimators $\mathcal{M} = \frac{1}{n}\sum_{i=1}^{n} X_i$ and $\mathcal{S}^2 = \frac{1}{n-1}\sum_{i=1}^{n}(X_i - \mathcal{M})^2$, they are independent, and that:

$$\frac{\mathcal{M} - \theta_0}{\sqrt{\frac{\mathcal{S}^2}{n}}} \sim t(n-1), \quad \frac{(n-1)\mathcal{S}^2}{\theta_1} \sim \chi^2(n-1) \tag{2.105}$$

2.4 Chi-squared test

Prove the basic result about the chi-squared test, that is that the Pearson random variable converges in distribution to a random variable $\chi^2(r-1)$, r being the number of intervals, as the rank of the sample tends to $+\infty$.

2.5 Velocity of light in the air

The Table 2.2 contains $n = 100$ measurements of the velocity of light in air by A. Michelson (1879): each value plus 299000 is a measure of c in km/s.

Estimate the mean and the variance, under the assumption that $X_i \sim N(\theta_0, \theta_1)$. Estimate confidence intervals for the mean and for the variance at the level $1 - \alpha = 95\%$. Use the Student test to test the hypothesis that the mean is equal to the exact value $c = 299792.458$ km/s.

Table 2.2 The 100 measurements of the velocity of light in air by A. Michelson (1879), from [4]; the given values plus 299000 re the original measurements in km/s

850	740	900	1070	930	850	950	980	980	880
1000	980	930	650	760	810	1000	1000	960	960
960	940	960	940	880	800	850	880	900	840
830	790	810	880	880	830	800	790	760	800
880	880	880	860	720	720	620	860	970	950
880	910	850	870	840	840	850	840	840	840
890	810	810	820	800	477	760	740	750	760
910	920	890	860	880	720	840	850	850	780
890	840	780	810	760	810	790	810	820	850
870	870	810	740	810	940	950	800	810	870

Table 2.3 Are these numbers drawn from a uniform distribution?

0.676636	0.231011	0.613735	0.055805
0.924277	0.335412	0.289339	0.927961
0.250062	0.809011	0.056113	0.661863
0.939963	0.966387	0.079119	0.759914
0.891149	0.554386	0.583501	0.912486

2.6 Kolmogorov-Smirnov and chi-squared test

Consider the set of data in Table 2.3:

Use Chi-squared with 5 bins of equal length and Kolmogorov-Smirnov tests to test the hypothesis at the confidence level 90% that such data can be modeled by a uniform law in $(0, 1)$.

2.7 Estimators for sums and products

Let $X_1 \ldots X_n$ and $Y_1 \ldots Y_n$ be independent samples with normally distributed components, such that $X_i \sim N(\mu_X, \sigma_X^2)$ and $Y_i \sim N(\mu_Y, \sigma_Y^2)$. Construct estimators for $\mu_X + \mu_Y$ and $\mu_X \mu_Y$.

References

1. Ross, Sheldon M.: Introductory Statistics. Elsevier Academic Press, Burlington (2005)
2. Cramer, H.: Mathematical Methods of Statistics. Princeton University Press (1999)
3. Cowan, G.: Statistical Data Analysis. Clarendon Press (1998)
4. Stigler, S.M.: Ann. Stat. **5**, 1055–1098 (1977)

Chapter 3
Conditional Probability and Conditional Expectation

Abstract In this chapter we deal with conditional probability. After having sketched the elementary definitions, we introduce the advanced notion of conditional expectation of a random variable with respect to a given σ-field. The intuitive meaning of the conditional expectation is the best prediction we can do about the values taken by the random variable, once we have observed the family of events inside the σ-field. The conditional expectation is widely used in the theory of stochastic processes we will present in the following chapters.

Keywords Conditional probability · Conditional expectation · Measurability Independence · Bayes theorem

3.1 Introduction

So far we have focused on independent random variables, which are well suitable to deal with statistical inference, allowing us to exploit the very powerful central limit theorem. However, for the purpose of studying time dependent random phenomena, it is necessary to consider a much wider class of random variables.

The treatment of mutually dependent random variables is based on the fundamental notion of *conditional probability* and on the more sophisticated *conditional expectation* which we will present in this chapter.

The conditional probability is presented in every textbook about probability; we sketch this topic briefly in the following section and then we turn to the conditional expectation, which will provide a very important tool to deal with stochastic processes.

3.2 Conditional Probability

Definition 3.1 Let $(\Omega, \mathcal{F}, P)$ be a probability space and $B \in \mathcal{F}$ an event with non-zero probability; the *conditional probability of A with respect to B* is:

© Springer International Publishing AG, part of Springer Nature 2018
E. Vitali et al., *Theory and Simulation of Random Phenomena*, UNITEXT
for Physics, https://doi.org/10.1007/978-3-319-90515-0_3

$$P(A|B) = \frac{P(A \cap B)}{P(B)} \tag{3.1}$$

Intuitively, $P(A|B)$ is the probability that the event A occurs when we know that the event B has occurred. It is very simple to show that the map:

$$A \in \mathcal{F}, \quad A \rightarrow P(A|B) \tag{3.2}$$

defines a new probability measure on $(\Omega, \mathcal{F})$. If the events A and B are independent, $P(A|B) = P(A)$: the occurring of the event B does not provide any information about the occurring of A.

From the Definition (3.1) of conditional probability the following important theorems easily follow:

Theorem 3.1 (Bayes) *Let* $(\Omega, \mathcal{F}, P)$ *a probability space and* $A, B \in \mathcal{F}$ *events,* $P(A) \neq 0$, $P(B) \neq 0$; *then:*

$$P(A|B)\,P(B) = P(B|A)\,P(A) \tag{3.3}$$

Proof Both members are equal to $P(A \cap B)$.

Theorem 3.2 (Law of Total Probability) *Let* $(\Omega, \mathcal{F}, P)$ *a probability space and* $A \in \mathcal{F}$ *an event and* $\{B_i\}_{i=1}^n$ *mutually disjoint events,* $P(B_i) \neq 0$, *such that* $\cup_i B_i = \Omega$; *then:*

$$P(A) = \sum_{i=1}^n P(A|B_i)P(B_i) \tag{3.4}$$

Proof

$$P(A) = \sum_{i=1}^n P(A \cap B_i) = \sum_{i=1}^n P(A|B_i)P(B_i)$$

3.3 Conditional Expectation

The remainder of the chapter is devoted to the presentation of the *conditional expectation*, which can be thought as a generalization of the concept of conditional probability: the idea of conditional expectation rises from the observation that the knowledge of one or more events, represented by a sub-σ-field $\mathcal{G} \subset \mathcal{F}$, allows to "predict" values taken by a random variable X through another random variable $E[X|\mathcal{G}]$, the so-called conditional expectation of X with respect to $\mathcal{G}$.

Before giving the rigorous definition of $E[X|\mathcal{G}]$, we discuss a useful example. Consider a sequence of random variables, taking values in $\mathbb{R}^n$, defined as:

$$\zeta_n = \sum_{i=1}^{n} \xi_i, \quad \zeta_0 = 0 \tag{3.5}$$

The label n could represent a time instant and ζ_n the position of a particle, resulting from the accumulation of n steps ξ_i. We will assume that $\{\xi_i\}$ are independent and identically distributed, with mean μ. We observe that, while $\{\xi_i\}$ are i.i.d, the random variables ζ_n and ζ_m, for $m \neq n$, are correlated. This correlation is crucial for describing time dependent phenomena, as it will become very clear in the following chapter.

Let us now denote $Y = \zeta_n$ and $X = \zeta_{n+k}$, for $k > 0$. Imagine that we can measure the position at time n, that is Y, while the time instant $n + k$ is in the future. Can we predict the position at time $n + k$, that is X, relying on our measurement of Y?

It is natural to proceed as follows: we try to build a function of Y, say $f(Y)$, that minimizes the intuitive figure of merit $E[(X - f(Y))^2]$. For simplicity, we will assume that the random variables $\{\xi_i\}$ are discrete, implying that Y and X are discrete. We can thus parametrize:

$$f(Y) = \sum_i \alpha_i 1_{Y=y_i} \tag{3.6}$$

where α_i is $f(y_i)$. Differentiating $E[(X - f(Y))^2]$ with respect to α_i we obtain the minimization condition:

$$\alpha_i = \frac{E[X 1_{Y=y_i}]}{E[1_{Y=y_i}]} = \sum_j x_j P(X = x_j | Y = y_i) \tag{3.7}$$

Now, we observe that:

$$P(X = x_j | Y = y_i) = \frac{P(X = x_j, Y = y_i)}{P(Y = y_i)} = P(\sum_{l=n+1}^{n+k} \xi_l = x_j - y_i) \tag{3.8}$$

so that:

$$\alpha_i = \sum_j x_j P(\sum_{l=n+1}^{n+k} \xi_l = x_j - y_i) = \sum_u (y_i + u) P(\sum_{l=n+1}^{n+k} \xi_l = u) = y_i + k\mu \tag{3.9}$$

We conclude thus that:

$$f(Y) = Y + k\mu \tag{3.10}$$

The interpretation is very simple: if we observe the position of the molecule at time n, the best prediction for the position of the particle at time $n + k$ is obtained drifting the current position by $k\mu$, μ being the average step.

In the rest of this chapter we will learn to call $f(Y)$ the conditional expectation of X given Y, $E[X|Y]$, which we will now define rigorously.

Let X be a real **integrable** random variable on a probability space $(\Omega, \mathcal{F}, P)$, and $\mathcal{G}$ a sub-σ-field of $\mathcal{F}$. Consider the map:

$$B \in \mathcal{G}, \quad B \rightarrow Q^{X,\mathcal{G}}(B) \stackrel{def}{=} \int_B X(\omega)P(d\omega) \tag{3.11}$$

If $X \geq 0$, (3.11) defines a positive measure on $(\Omega, \mathcal{G})$, absolutely continuous with respect to P; by virtue of the Radon-Nikodym theorem [1], there exists a real random variable Z, $\mathcal{G}$-**measurable**, a.s. unique and such that:

$$Q^{X,\mathcal{G}}(B) = \int_B Z(\omega)P(d\omega) \quad \forall B \in \mathcal{G} \tag{3.12}$$

Such random variable will be denoted:

$$Z = E[X|\mathcal{G}] \tag{3.13}$$

and called **conditional expectation** of X given $\mathcal{G}$. If X is not positive, it can be represented as difference of two positive random variables $X = X^+ - X^-$ and its conditional expectation given $\mathcal{G}$ can be defined as follows:

$$Z = E[X|\mathcal{G}] = E\left[X^+|\mathcal{G}\right] - E\left[X^-|\mathcal{G}\right] \tag{3.14}$$

The conditional expectation $E[X|\mathcal{G}]$ is defined by the two conditions:
1. $E[X|\mathcal{G}]$ is $\mathcal{G}$-measurable
2. $E[1_B E[X|\mathcal{G}]] = E[1_B X]$, $\quad \forall B \in \mathcal{G}$
   With measure theory arguments it can be proved that the second condition is equivalent to:
   $$E[W E[X|\mathcal{G}]] = E[W X] \tag{3.15}$$

for all **bounded** and $\mathcal{G}$-**measurable** random variables W.

The key point is the $\mathcal{G}$-**measurability**: if $\mathcal{G}$ represents the amount of information available, in general we cannot access all information about X; on the other hand, we can construct $E[X|\mathcal{G}]$, whose distribution is known, and use it to replace X whenever events belonging to $\mathcal{G}$ are considered.

We will often use the notation:

$$P(A|\mathcal{G}) \stackrel{def}{=} E[1_A|\mathcal{G}] \tag{3.16}$$

and call (3.16) the **conditional probability** of A given $\mathcal{G}$; we stress that (3.16), in contrast to (3.1), is a random variable. Moreover, it is $\mathcal{G}$-measurable and such that:

$$\int_B P(A|\mathcal{G})(\omega)\,P(d\omega) = \int_B 1_A(\omega)P(d\omega) = P(A \cap B), \quad \forall B \in \mathcal{G} \qquad (3.17)$$

3.4 An Elementary Construction

We will now provide some intuitive and practical insight into the formal definition of conditional expectation, giving an elementary construction of (3.12) under some simplifying hypotheses. Let $Y : \Omega \to \mathbb{R}$ be a random variable and, as usual:

$$\sigma(Y) \stackrel{def}{=} \left\{ A \subset \Omega \,|\, A = Y^{-1}(B),\, B \in \mathcal{B}(\mathbb{R}) \right\} \qquad (3.18)$$

the σ-field **generated by** Y, i.e. the smallest sub-σ-field of $\mathcal{F}$ with respect to which Y is measurable. Intuitively, if our amount of information is $\sigma(Y)$, this means that, after an experiment, we know only the value of Y: we do not have access to other information.

Let us assume that Y be **discrete**, i.e. that it can assume at most countably infinite values $\{y_1, \ldots, y_n, \ldots\}$. For all events $A \in \mathcal{F}$, let us define:

$$P(A|Y = y_i) \stackrel{def}{=} \begin{cases} \frac{P(A \cap \{Y=y_i\})}{P(Y=y_i)} & \text{if } P(Y = y_i) > 0 \\ 0 & otherwise \end{cases} \qquad (3.19)$$

Equation (3.19) is nothing but the familiar conditional probability of A given the event $\{Y = y_i\}$. Due to the law of total probability:

$$P(A) = \sum_i P(A|Y = y_i)\,P(Y = y_i) \qquad (3.20)$$

Let now $X : \Omega \to \mathbb{R}$ be an integrable random variable, which we assume **discrete** for simplicity, and consider the map:

$$\mathcal{B}(\mathbb{R}) \ni H \to P(X \in H|Y = y_i) \qquad (3.21)$$

Equation (3.21) is a probability measure on $\mathbb{R}$, with expectation:

$$E[X|Y = y_i] \stackrel{def}{=} \sum_j x_j P\left(X = x_j|Y = y_i\right) \qquad (3.22)$$

Consider now the function $h : \mathbb{R} \to \mathbb{R}$:

$$y \in \mathbb{R} \ \rightarrow \ h(y) \overset{def}{=} \begin{cases} E\,[X|Y = y_i], & \text{if } y = y_i, \ \ P\,(Y = y_i) > 0 \\ 0 & otherwise \end{cases} \tag{3.23}$$

h is clearly measurable, so that it makes perfectly sense to construct the random variable $Z : \Omega \rightarrow \mathbb{R}$

$$\omega \in \Omega \mapsto Z(\omega) \overset{def}{=} h\,(Y(\omega)) \tag{3.24}$$

that is, recalling (3.22):

$$Z(\omega) = \begin{cases} E\,[X|Y = y_i], & \text{if } Y(\omega) = y_i, \ \ P\,(Y = y_i) > 0 \\ arbitrary\,value & otherwise \end{cases} \tag{3.25}$$

Equation (3.25) has a straightforward interpretation: given a realization of Y, the expectation of the possible outcomes of X can be computed; this expectation is random, like Y. At a first sight, the arbitrary constant in (3.25) could seem disturbing: nevertheless, the subset of Ω on which Z takes an arbitrary value has probability equal to 0.

Incidentally, we remark that, for all $A \in \mathcal{F}$:

$$E\,[1_A|Y = y_i] = \sum_{a=0,1} a\,P\,(1_A = a|Y = y_i) = P\,(A|Y = y_i) \tag{3.26}$$

It remains to show that:

$$Z(\omega) = E\,[X|\sigma(Y)]\,(\omega) \quad a.s. \tag{3.27}$$

First, we show that $Z : \Omega \rightarrow \mathbb{R}$ is $\sigma(Y)$-measurable; to this purpose, consider $H \in \mathcal{B}(\mathbb{R})$:

$$Z^{-1}(H) = (h \circ Y)^{-1}\,(H) = Y^{-1}\left(h^{-1}(H)\right) \in \sigma(Y) \tag{3.28}$$

since $h^{-1}(H) \in \mathcal{B}(\mathbb{R})$. Therefore, Z is $\sigma(Y)$-measurable. Let now be W a bounded and $\sigma(Y)$-measurable random variable (this includes the case $W = 1_B$, with $B \in \sigma(Y)$). By virtue of a theorem by J. L. Doob, which we state without proof reminding the interesting reader to [2], there exists a measurable function $w : \mathbb{R} \rightarrow \mathbb{R}$ such that $W = w(Y)$. Recalling that $Z = h(Y)$, we therefore have:

$$\begin{aligned} E\,[WZ] &= \sum_i w(y_i)E\,[X|Y = y_i]\,P(Y = y_i) = \\ &= \sum_i w(y_i) \sum_j x_j P\,(X = x_j|Y = y_i)\,P(Y = y_i) = \\ &= \sum_{i,j} w(y_i)x_j P\,(\{X = x_j\} \cap \{Y = y_i\}) = E\,[WX] \end{aligned} \tag{3.29}$$

which is exactly the second condition defining $E\,[X|\sigma(Y)]$.

3.5 Computing Conditional Expectations from Probability Densities

The notion of conditional expectation $E[X|\sigma(Y)]$ is well defined also for continuous random variables X, Y. Several authors write, for the sake of simplicity, $E[X|Y]$ instead of $E[X|\sigma(Y)]$. Remarkably, since $E[X|\sigma(Y)]$ is $\sigma(Y)$-measurable, by virtue of Doob's theorem [2] there exists a measurable function g such that:

$$E[X|Y] = g(Y) \qquad (3.30)$$

Equation (3.30) has the intuitive interpretation that to predict X given Y it is sufficient to apply a measurable "deterministic" function to Y. In the remainder of this section, we will present a practical way to compute explicitly $g(Y)$ in some special situations. The following discussion will be based on the simplifying assumptions:

1. that the random variables X, Y take values in $\mathbb{R}$.
2. that X and Y have joint law absolutely continuous with respect to the Lebesgue measure, and therefore admit joint probability density $p(x, y)$.
3. that the joint probability density $p(x, y)$ is a.e. non-zero.

Under these hypotheses, the marginal probability densities and the conditional probability density of X given Y can be defined with the formulas:

$$p_X(x) = \int dy \, p(x, y) \qquad p_Y(y) = \int dx \, p(x, y) \qquad p(x|y) = \frac{p(x, y)}{p_Y(y)} \quad (3.31)$$

By virtue of these definitions, and of Fubini's theorem [1], we find that for all bounded measurable functions $h : \mathbb{R} \to \mathbb{R}$ one has:

$$E[Xh(Y)] = \int dx \int dy \, p(x, y) \, x \, h(y) =$$
$$= \int dy \, p_Y(y) \, h(y) \int dx \, x \, p(x|y) = E[g(Y)h(Y)]$$

where the measurable function:

$$g(y) = \int dx \, x \, p(x|y)$$

has appeared. Since, on the other hand:

$$E[Xh(Y)] = E[E[X|Y]h(Y)]$$

we conclude that $E[X|Y] = g(Y)$. As an example of remarkable importance, consider a bivariate normal random variable $(X, Y) \sim N(\mu, \Sigma)$ with joint density:

$$p(x, y) = \frac{e^{-\frac{(x-\mu_x)^2}{2\sigma_x^2(1-\rho^2)} + \frac{\rho(x-\mu_x)(y-\mu_y)}{\sigma_x\sigma_y(1-\rho^2)} - \frac{(y-\mu_y)^2}{2\sigma_y^2(1-\rho^2)}}}{2\pi\sigma_x\sigma_y\sqrt{1-\rho^2}}$$

A straightforward calculation shows that:

$$p_Y(y) = \frac{e^{-\frac{(y-\mu_y)^2}{2\sigma_y^2}}}{\sqrt{2\pi\sigma_y^2}}$$

which allows us to obtain $p(x|y)$ and to compute explicitly $g(y)$. The result is:

$$g(y) = \mu_x + \frac{\rho\,\sigma_x}{\sigma_y}(y - \mu_y)$$

The conditional expectation of X given Y is therefore, in this special situation, *a linear function of Y*, explicitly depending on the elements of the covariance matrix Σ.

3.6 Properties of Conditional Expectation

The following theorem contains some important properties of the conditional expectation.

Theorem 3.3 *Let X be a real integrable random variable, defined on a probability space $(\Omega, \mathcal{F}, P)$, and $\mathcal{G}$ a sub-σ-field of $\mathcal{F}$. Then:*

1. *the map $X \to E[X|\mathcal{G}]$ is a.s. linear*
2. *if $X \geq 0$ a.s., then $E[X|\mathcal{G}] \geq 0$ a.s.*
3. *$E[E[X|\mathcal{G}]] = E[X]$.*
4. *if X is $\mathcal{G}$-measurable, then $E[X|\mathcal{G}] = X$ a.s.*
5. *if X is independent on $\mathcal{G}$, i.e. is if $\sigma(X)$ and $\mathcal{G}$ are independent, then $E[X|\mathcal{G}] = E[X]$ a.s.*
6. *if $\mathcal{H} \subset \mathcal{G}$ is a σ-field, then $E[E[X|\mathcal{G}]|\mathcal{H}] = E[X|\mathcal{H}]$ a.s.*
7. *if Y is bounded and $\mathcal{G}$-measurable, then $E[YX|\mathcal{G}] = YE[X|\mathcal{G}]$ a.s.*

Proof The first two points are obvious. To prove the third one, it is sufficient to recall that:

$$E[1_B E[X|\mathcal{G}]] = E[1_B X], \quad \forall B \in \mathcal{G} \tag{3.32}$$

and choose $B = \Omega$. To prove the fourth point, it is sufficient to observe that, in such case, X itself satisfies the two conditions defining the conditional expectation.

To prove the fifth point, we observe that, since the random variable $\omega \mapsto E[X]$ is constant and therefore $\mathcal{G}$-measurable, and since for all $B \in \mathcal{G}$ the random variable 1_B is clearly $\mathcal{G}$-measurable and independent on X:

$$E[1_B X] = E[1_B] E[X] = E[1_B E[X]] \tag{3.33}$$

so that $\omega \mapsto E[X]$ satisfies the two conditions defining the conditional expectation.

To prove the sixth point, we observe that by definition $E[E[X|\mathcal{G}]|\mathcal{H}]$ is $\mathcal{H}$-measurable; moreover, since $B \in \mathcal{H}$, 1_B is $\mathcal{H}$-measurable and also $\mathcal{G}$-measurable (since $\mathcal{G}$ contains $\mathcal{H}$). Therefore:

$$E[1_B E[E[X|\mathcal{G}]|\mathcal{H}]] = E[1_B E[X|\mathcal{G}]] = E[1_B X] \tag{3.34}$$

where the definition of conditional expectation has been applied twice.

To prove the last point we observe that, as a product of $\mathcal{G}$-measurable random variables, $YE[X|\mathcal{G}]$ is $\mathcal{G}$-measurable. For all bounded, $\mathcal{G}$-measurable random variables Z, therefore:

$$E[Z\,YE[X|\mathcal{G}]] = E[Z\,YX] = E[Z\,E[YX|\mathcal{G}]] \tag{3.35}$$

since also ZY is bounded and $\mathcal{G}$-measurable.

3.7 Conditional Expectation as Prediction

We are going now to put on a firm ground the intuitive idea of the conditional expectation as prediction: $E[X|\mathcal{G}]$ "predicts" X when the amount of information is $\mathcal{G}$. To this purpose, we need a geometrical interpretation: let $L^2(\Omega, \mathcal{F}, P)$ be the Hilbert space of (complex valued) square-integrable random variables, endowed with the inner product:

$$\langle X|Y \rangle \overset{def}{=} E[\overline{X}Y] \tag{3.36}$$

Moreover, as usual $\mathcal{G}$ is sub-σ-field of $\mathcal{F}$. The space $L^2(\Omega, \mathcal{G}, P)$ is a closed subspace $L^2(\Omega, \mathcal{F}, P)$. Let us define the mapping:

$$X \in L^2(\Omega, \mathcal{F}, P), \quad X \to \hat{Q}X \overset{def}{=} E[X|\mathcal{G}] \tag{3.37}$$

We leave to the reader the simple proof of the fact that $\hat{Q}$ is a **linear operator** from $L^2(\Omega, \mathcal{F}, P)$ to $L^2(\Omega, \mathcal{G}, P)$.

Let us prove that $\hat{Q}$ is idempotent, that is $\hat{Q}^2 = \hat{Q}$:

$$\begin{aligned}
\hat{Q}^2 X &= \hat{Q}E[X|\mathcal{G}] = E[E[X|\mathcal{G}]|\mathcal{G}] = \\
&= E[X|\mathcal{G}] = \hat{Q}X
\end{aligned} \tag{3.38}$$

Moreover $\hat{Q}$ is self-adjoint, since:

$$\langle X | \hat{Q} Y \rangle = E \left[\overline{X} E [Y | \mathcal{G}] \right] = E \left[E \left[\overline{X} E [Y | \mathcal{G}] | \mathcal{G} \right] \right] =$$
$$= E \left[E [Y | \mathcal{G}] E \left[\overline{X} | \mathcal{G} \right] \right] = E \left[E \left[\overline{Y} E \left[\overline{X} | \mathcal{G} \right] | \mathcal{G} \right] \right] = \qquad (3.39)$$
$$= E \left[Y E \left[\overline{X} | \mathcal{G} \right] \right] = E \left[E \left[\overline{X} | \mathcal{G} \right] Y \right] = \langle \hat{Q} X | Y \rangle$$

Therefore $\hat{Q}$ is an **orthogonal projector** onto the subspace $L^2 (\Omega, \mathcal{G}, P)$.

Let now be $X \in L^2 (\Omega, \mathcal{F}, P)$ a real random variable; let us look for the element $Y \in L^2 (\Omega, \mathcal{G}, P)$ such that $||X - Y||^2$ is minimum. The minimum is reached for $Y = E [X | \mathcal{G}]$. The key point is that $Y = \hat{Q} Y$, in fact:

$$E \left[(X - Y)^2 \right] = ||X - Y||^2 = ||\hat{Q} X + (1 - \hat{Q}) X - \hat{Q} Y||^2 =$$
$$= ||\hat{Q} (X - Y)||^2 + ||(1 - \hat{Q}) X||^2 = ||Y - \hat{Q} X||^2 + ||(1 - \hat{Q}) X||^2 \qquad (3.40)$$

and the minimum is achieved precisely at $Y = \hat{Q} X$.

In the sense of L^2, thus, $Y = E [X | \mathcal{G}]$ is the best approximation of X among the class of $\mathcal{G}$-measurable functions. This is the justification of the interpretation of $Y = E [X | \mathcal{G}]$ as a prediction: within the set of square-integrable $\mathcal{G}$-measurable random variables, $Y = E [X | \mathcal{G}]$ is the closest one to X in the topology of L^2.

3.8 Linear Regression and Conditional Expectation

There is a very interesting connection between conditional expectation and linear regression which we are going now to explore. Let's consider two real random variables Y and Z, representing two properties one wishes to measure during an experiment. Quite often it happens that the quantity Z can be measured with an high accuracy, while Y, a "response", contains a signal and a noise difficult to disentangle. In such situations, from a mathematical point of view, the experimentalist would like to work with $\sigma (Z)$-measurable random variables: such quantities, in fact, have a well defined value once the outcome of Z is known. The key point is that $E[Y|Z]$ is the best prediction for Z within the set of $\sigma (Z)$-measurable random variables.

Since the conditional expectation is a linear projector, we can always write the unique decomposition:

$$Y = E[Y|Z] + \varepsilon \qquad (3.41)$$

where ε is a real random variable. It is immediate to show that:

$$E[\varepsilon] = E[\varepsilon|Z] = 0 \qquad (3.42)$$

Moreover, since $Y - E[Y|Z]$ is orthogonal to all the random variables $\sigma (Z)$-measurable:

$$E[(Y - E[Y|Z])h(Z)] = 0 \qquad (3.43)$$

in particular the following orthogonality property holds:

$$E[\varepsilon Z] = 0 \tag{3.44}$$

In order to proceed further, let's assume that the **two-dimensional random variable** (Z, X) **is normal**. In such case we know that the conditional expectation depends linearly on Z:

$$E[Y|Z] = a + bZ \tag{3.45}$$

with:

$$
\begin{aligned}
a &= \frac{Var(Z)E[Y] - E[Z]Cov(Z, Y)}{Var(Z)} \\
b &= \frac{Cov(Z, Y)}{Var(Z)}
\end{aligned}
\tag{3.46}
$$

The "error" $\varepsilon = Y - (a + bZ)$ is also normal being a linear function of (X, Z) of zero mean. Moreover, since $E[\varepsilon Z] = 0$, ε is **independent** of Z. The variance is:

$$\sigma^2 = Var(\epsilon) = E\left[(Y - (a + bZ))^2\right] \tag{3.47}$$

and, from the geometrical interpretation of the conditional expectatio, we know that the parameters in (3.46) minimize such quantity.

To summarize, we have found that, whenever two quantities Z and Y have a joint normal law, the best prediction we can do for Y once we know the value of Z is a linear function of such value. The experimentalist collects a set of data $\{(z_1, y_1), \ldots, (z_n, y_n)\}$, interprets such data as realization of two-dimensional random variables (Z_i, Y_i) independent and identically distributed as (Z, Y), and uses such data to infer the values of a, b and σ^2, the last one providing the "accuracy" of the linear approximation.

For the statistical analysis, we refer to the previous chapter.

3.9 Conditional Expectation, Measurability and Independence

The conclusion of the present chapter is devoted to the presentation of a useful result, which will be used later. For clarity, let us think about a particle moving from an initial position X to a final position $X + Y$. We assume that X is $\mathcal{G}$-measurable, meaning that $\mathcal{G}$ contains all the necessary information to know the initial position, and that Y is independent from $\mathcal{G}$. What the best prediction for the final position, given the initial one? We learnt in this chapter that the best prediction is:

$$E[X + Y|\mathcal{G}] = X + E[Y] \tag{3.48}$$

where we have used the properties of conditional expectation. The result is a random variable, function of X.

More generally, what is the best prediction for a function (let's imagine for example the value of a field), evaluated in the final position, or any function of both the initial and the final position (imagine the calculation of the velocity)? That is, we wish to compute $E[g(X, Y)|\mathcal{G}]$ for a given function g. Intuitively, we expect that the result will be a random variable function of X. We will now show how to make the explicit calculation.

In the simple case $g(X, Y) = XY$, we have:

$$E[XY|\mathcal{G}] = XE[Y|\mathcal{G}] = XE[Y] \tag{3.49}$$

This can be immediately generalized to linear combinations of factorized functions $g(X, Y) = f(X)h(Y), f, h$ being measurable functions (and integrability conditions have naturally to be fulfilled).

$$E[f(X)h(Y)|\mathcal{G}] = f(X)E[h(Y)|\mathcal{G}] = f(X)E[h(Y)] \tag{3.50}$$

The key point is that the result is a random variable depending on X, whose explicit form is obtained performing an expectation over Y. Formally, we can summarize this result in the following theorem [3], where we introduce a function $\psi(x, \omega)$ such that:

$$g(X(\omega), Y(\omega)) = \psi(x, \omega)|_{x=X(\omega)} \tag{3.51}$$

Theorem 3.4 *Let $(\Omega, \mathcal{F}, P)$ be a probability space, $\mathcal{G}$ and $\mathcal{H}$ **mutually independent** sub-σ-fields of $\mathcal{F}$. let $X : \Omega \to E$ be a $\mathcal{G}$-**measurable** random variable taking values in the measurable space $(E, \mathcal{E})$ and ψ a function $\psi : E \times \Omega \to \mathbb{R}$ $\mathcal{E} \otimes \mathcal{H}$-**measurable**, such that $\omega \mapsto \psi(X(\omega), \omega)$ is integrable. Then:*

$$E[\psi(X, \cdot)|\mathcal{G}] = \Phi(X), \quad \Phi(x) \stackrel{def}{=} E[\psi(x, \cdot)] \tag{3.52}$$

3.10   Further Readings

Readers wishing to deepen their knowledge about conditional probability and conditional expectation can refer to many excellent textbooks, like, e.g., [4, 5].

Problems

3.1 Random summations
Let $\{X\}_i$ be a sequence of independent and identically distributed, taking valued in $\mathbb{N}$. Let, moreover, N be another random variable taking values in $\mathbb{N}$, independent from the X_i. Define the random summation:

$$S_N = X_1 + \cdots + X_N \tag{3.53}$$

Evaluate the discrete density $P(S_N = k)$ and the **generating function**:

$$\psi_{S_N}(z) = E[z^{S_N}] \tag{3.54}$$

Deduce the useful relation:

$$E[S_N] = E[N]E[X_i] \tag{3.55}$$

which holds provided that N and X_i are integrable.

3.2 Lack of memory

Let X be a geometric random variable with parameter p, i.e. with density

$$p(x) = \begin{cases} p(1-p)^x & x = 0, 1, 2, \ldots \\ 0 & otherwise \end{cases} \tag{3.56}$$

Show that:

$$P(X \geq j + k \,|\, X \geq j) = P(X \geq k) \tag{3.57}$$

Show that the same is true if X is exponential with parameter λ, i.e. with density:

$$p(x) = \lambda \exp(-\lambda x)\, 1_{(0,+\infty)}(x) \tag{3.58}$$

Precisely, show that:

$$P(X \geq t + s \,|\, X \geq t) = P(X \geq s) \tag{3.59}$$

3.3 Diagnostic test

In medicine, a diagnostic test is any kind of medical test to aid in the diagnosis of a disease. Suppose that, if the patient has the disease, the probability that the test is positive is 99% and, contemporarily, the probability that the test if negative assuming that the patient does not have the disease is the same 99%. Now, assuming that the incidence of the given disease is 0.2%, what is the probability that any individual whose test has turned out to be positive actually has the disease?

3.4 A simple conditional expectation

Show that, in the special case $\mathcal{G} = \{\emptyset, \Omega, A, A^C\}$ where $A \in \mathcal{F}$ has non-zero probability, we have:

$$E[X \,|\, \mathcal{G}](\omega) = \begin{cases} \frac{1}{P(A)} \int_A P(d\omega) X(\omega), & \omega \in A \\ \frac{1}{P(A^C)} \int_{A^C} P(d\omega) X(\omega), & \omega \in A^C \end{cases} \tag{3.60}$$

3.5 Calculation of conditional expectation

Consider a random variable (X, Y) uniform inside the unit circle, with density:

$$p_{(X,Y)}(\mathbf{x}) = \begin{cases} \frac{1}{\pi}, & |\mathbf{x}| \leq 1 \\ 0, & otherwise \end{cases} \tag{3.61}$$

Find $E[X|Y]$.

3.6 A simple situation

Suppose a random variable X is used to model the measurement of a quantity which is assumed to be normal with mean M and variance 1; the mean M is random too and is assumed to follow an exponential law with parameter λ. Evaluate $E[X]$. What is the joint probability density $p(x, m)$ of (X, M)?

3.7 Gamma and negative binomial law

We say that a random variable X follows a Gamma law with parameters $\alpha, \beta, \alpha > 0$ and $\beta > 0$, and write $X \sim \Gamma(\alpha, \beta)$, if its density is:

$$p_X(x) = \frac{\beta^\alpha}{\Gamma(\alpha)} x^{\alpha-1} \exp(-\beta x) \, 1_{(0,+\infty)}(x) \tag{3.62}$$

We observe that $\Gamma(1, \beta)$ is the exponential law with parameter β, while $\Gamma\left(\frac{n}{2}, \frac{1}{2}\right) = \chi^2(n)$.

Now, suppose a random variable N represents a quantity that is modeled by a Poisson distribution with a random parameter $\Lambda \sim \Gamma(\alpha, \beta)$. Show that N follows a negative binomial law with parameters $\alpha > 0$ and $0 \leq p = \frac{\beta}{\beta+1} \leq 1$, that is:

$$p_N(n) = \begin{cases} \frac{(\alpha+n-1)(\alpha+n-2)...\alpha}{n!} \, p^\alpha \, (1-p)^n, & n = 0, 1, 2, \ldots \\ 0, & otherwise \end{cases} \tag{3.63}$$

Evaluate $E[N]$.

References

1. Rudin, W.: Real and Complex Analysis. McGraw-Hill (1987)
2. Rao, M.M., Swift, R.J.: Probability Theory with Applications. Springer US (2006)
3. Baldi, P.: Equazioni Differenziali Stocastiche e Applicazioni. Pitagora Editrice (2000)
4. Feller, W.: An Introduction to Probability Theory and its Applications. Wiley (1966)
5. Kallenberg, O.: Foundations of Modern Probability. Springer NY (2001)

Chapter 4
Markov Chains

Abstract In this chapter we will start dealing with stochastic processes, which are the mathematical models for phenomena whose temporal evolution contains some randomness. Starting from the celebrated example of the random walk, we will introduce the central definition of Markov chains, which, although simple, provide extremely important models for physical systems. The description of Markov chains will allow us to introduce the central topic of *thermalization* and approach to equilibrium of random motions, that is the existence of asymptotic laws to which the Markov chains converge, in a sense that will be made rigorous. Finally, we will introduce Metropolis theorem, a cornerstone of numerical simulations, as will be discussed in the following chapter.

Keywords Markov chains · Random walk · Transition matrix · Invariant laws Metropolis theorem

4.1 Basic Definitions

In the present chapter we will introduce the mathematical description of time-dependent random phenomena. We will begin treating the simple case in which the time evolution can be represented as a sequence of steps in discrete time, and the random variables describing the quantities evolving randomly in discrete time take values in a discrete space.

In the following, we will consider a probability space $(\Omega, \mathcal{F}, P)$ and a set E **at most countable**, which we will call **state space**.

All the random variables X which will be dealt with are measurable functions $X : \Omega \to E$ with discrete density:

$$k \in E \to v_k \overset{def}{=} P(X = k), \quad v_k \geq 0, \quad \sum_{k \in E} v_k = 1 \tag{4.1}$$

As discussed in the first chapter, a discrete density uniquely defines a law: having in mind (4.1), for the sake of simplicity, we will call v the law of X with innocuous abuse of notation.

Random processes with discrete time and discrete state space can be interpreted as random walks on the points of E. To the purpose of describing such processes we must know the *transition probability* from a generic point $k \in E$ to another one. Therefore, a central ingredient in our treatment is represented by the following:

Definition 4.1 A **transition matrix** $\mathcal{P}$ on E is a real matrix, satisfying the following properties:

1. $\forall i, j \in E \quad 0 \leq \mathcal{P}_{i \to j} \leq 1$
2. $\forall i \in E \quad \sum_{j \in E} \mathcal{P}_{i \to j} = 1$

where the symbol $\mathcal{P}_{i \to j}$, $i, j \in E$ denotes the matrix elements of $\mathcal{P}$.

The above requirements enable us to interpret $\mathcal{P}_{i \to j}$ as the probability of moving from $i \in E$ to $j \in E$ in one time step: the second one, in particular, simply means that the probability of transitioning from i to any state in E is equal to one. We now give the fundamental:

Definition 4.2 Given a law v on E and a transition matrix $\mathcal{P}$ on E, we call **homogeneous Markov chain** with sample space E, with initial law v and transition matrix $\mathcal{P}$, a family: $\{X_n\}_{n \geq 0}$ of random variables $X_n : \Omega \to E$ such that:

1. X_0 has law v
2. whenever conditional probabilities make sense:

$$P(X_{n+1} = j | X_n = i, X_{n-1} = i_{n-1}, \ldots, X_0 = i_0) = \qquad (4.2)$$
$$= P(X_{n+1} = j | X_n = i) = \mathcal{P}_{i \to j}$$

We stress that $P(X_{n+1} = j | X_n = i)$ is assumed independent of n, whence the adjective **homogeneous** in definition (4.2).

4.2 Random Walk in d Dimensions

We begin our exposition of the theory of Markov chains with a remarkable example, the random walk on a lattice, which is the mathematical model of a path that consists of a succession of random steps. The reader can think of a drunk man, or a particle randomly moving, who, at each time, chooses a random direction, and makes a step accordingly.

In order to introduce the model, we let $\{\mathbf{e}_\mu\}_{\mu=1}^d$ be the canonical basis of $\mathbb{R}^d$,

$$\{\mathbf{e}_\mu\}_\mu = \{(1, 0, \ldots, 0), (0, 1, \ldots, 0), \ldots, (0, 0, \ldots, 1)\} \tag{4.3}$$

and we consider a family of independent and identically distributed random variables, $\{\xi_m\}_{m\in\mathbb{N}}$ taking values in the set of the $2d$ unit vectors (the directions) $\{\pm\mathbf{e}_\mu\}_\mu$. We assume that there is no preferred direction, that is:

$$P(\xi_m = \pm\mathbf{e}_\mu) = \frac{1}{2d} \tag{4.4}$$

Finally, we define:

$$X_0 = \mathbf{0}, \quad X_n = \xi_1 + \xi_2 + \cdots + \xi_n \tag{4.5}$$

X_n has the natural interpretation of position, at time n, of the walker starting from the origin $\mathbf{0}$ and moving, at each time step, choosing randomly a direction leading him/her to one of the nearest neighbors of its actual position in $\mathbb{Z}^d$.

With this position, we have defined an homogeneous Markov chain with initial law v, $\mathbf{x} \to v_{\mathbf{x}} = \delta_{\mathbf{x},0}$ and transition matrix:

$$\mathcal{P}_{\mathbf{x}\to\mathbf{y}} = \begin{cases} \frac{1}{2d}, & |\mathbf{x} - \mathbf{y}| = 1 \\ 0, & otherwise \end{cases} \tag{4.6}$$

In fact, choosing by construction $\mathbf{x}_0 = 0$ and $\mathbf{x}, \mathbf{x}_{n-1}, \ldots, \mathbf{x}_1$ nearest neighbors, we have:

$$P(X_{n+1} = \mathbf{y}|X_n = \mathbf{x}, X_{n-1} = \mathbf{x}_{n-1}, \ldots, X_0 = 0) =$$
$$= P(X_n + \xi_{n+1} = \mathbf{y}|X_n = \mathbf{x}, X_{n-1} = \mathbf{x}_{n-1}, \ldots, X_0 = 0) =$$
$$= \frac{P(X_n + \xi_{n+1} = \mathbf{y}, X_n = \mathbf{x}, X_{n-1} = \mathbf{x}_{n-1}, \ldots, X_0 = 0)}{P(X_n = \mathbf{x}, X_{n-1} = \mathbf{x}_{n-1}, \ldots, X_0 = 0)} = \tag{4.7}$$
$$= \frac{P(\xi_{n+1} = \mathbf{y} - \mathbf{x}, \xi_n = \mathbf{x} - \mathbf{x}_{n-1} \ldots, X_0 = 0)}{P(\xi_n = \mathbf{x} - \mathbf{x}_{n-1} \ldots, X_0 = 0)} =$$
$$= P(\xi_{n+1} = \mathbf{y} - \mathbf{x}) = \mathcal{P}_{\mathbf{x}\to\mathbf{y}}$$

where the fact that ξ_{n+1} is independent on all ξ_m with $m \le n$ has been used.

4.2.1 An Exact Expression for the Law

We start from the basic relation:

$$P(X_n = \mathbf{x}) = \sum_{\mathbf{y}\in\mathbb{Z}^d} P(X_{n-1} = \mathbf{y}) \, \mathcal{P}_{\mathbf{y}\to\mathbf{x}} = \frac{1}{2d} \sum_{\mathbf{y}\in\mathbb{Z}^d, |\mathbf{y}-\mathbf{x}|=1} P(X_{n-1} = \mathbf{y}) \tag{4.8}$$

Since the law at time n, $\mathbf{x} \rightarrow P(X_n = \mathbf{x})$, is defined on the lattice $\mathbb{Z}^d$, we can always turn in Fourier space, writing:

$$P(X_n = \mathbf{x}) = \frac{1}{(2\pi)^d} \int_{[-\pi,\pi]^d} d\mathbf{k} \; \mathcal{C}(n, \mathbf{k}) \exp(-i\mathbf{k} \cdot \mathbf{x}) \qquad (4.9)$$

We already know that:

$$\mathcal{C}(n = 0, \mathbf{k}) = 1 \qquad (4.10)$$

If we plug (4.9) into (4.8) we obtain:

$$\int_{[-\pi,\pi]^d} d\mathbf{k} \; \mathcal{C}(n, \mathbf{k}) \exp(-i\mathbf{k} \cdot \mathbf{x}) =$$

$$= \frac{1}{2d} \int_{[-\pi,\pi]^d} d\mathbf{k} \; \mathcal{C}(n-1, \mathbf{k}) \sum_{\mu=1}^{d} \left(\exp\left(-i\mathbf{k} \cdot (\mathbf{x} + \mathbf{e}_\mu)\right) + \exp\left(-i\mathbf{k} \cdot (\mathbf{x} - \mathbf{e}_\mu)\right) \right)$$

$$(4.11)$$

Implying that:

$$\mathcal{C}(n, \mathbf{k}) = \mathcal{C}(n-1, \mathbf{k}) \left(\frac{1}{d} \sum_{\mu=1}^{d} \cos(k_\mu) \right) \qquad (4.12)$$

Iterating we readily obtain:

$$\mathcal{C}(n, \mathbf{k}) = \left(\frac{1}{d} \sum_{\mu=1}^{d} \cos(k_\mu) \right)^n \qquad (4.13)$$

implying that:

$$P(X_n = \mathbf{x}) = \frac{1}{(2\pi)^d} \int_{[-\pi,\pi]^d} d\mathbf{k} \; \left(\frac{1}{d} \sum_{\mu=1}^{d} \cos(k_\mu) \right)^n \exp(-i\mathbf{k} \cdot \mathbf{x}) \qquad (4.14)$$

This is an exact expression for the law of the random walk at the time instant n.

4.2.2 Explicit Expression in One Dimension and Heat Equation

In the special case of the random walk in one dimension, we can perform the integral analytically:

$$P\left(X_n = x\right) = \frac{1}{(2\pi)} \int_{-\pi}^{\pi} dk \ (\cos(k))^n \exp\left(-ik \cdot x\right) =$$

$$= \frac{1}{(2\pi)} \int_{-\pi}^{\pi} dk \ \left(\frac{\exp(ik) + \exp(-ik)}{2}\right)^n \exp\left(-ik \cdot x\right) =$$

$$= \frac{1}{2^n} \frac{1}{(2\pi)} \int_{-\pi}^{\pi} dk \ \sum_{p=1}^{n} \binom{n}{p} \exp(ikp) \exp(-ik(n-p)) \exp\left(-ik \cdot x\right) = \qquad (4.15)$$

$$= \frac{1}{2^n} \binom{n}{\frac{1}{2}(n+x)}$$

provided that $\frac{1}{2}(n+x) \in \{0, 1, \ldots, n\}$.

We take the opportunity to observe that:

$$P(X_{n+1} = j) = P\left(\bigcup_{h \in \mathbb{Z}} \{X_{n+1} = j, X_n = h\}\right) =$$

$$= \sum_{h \in \mathbb{Z}} P(X_{n+1} = j, X_n = h) = \sum_{h \in \mathbb{Z}} P(X_{n+1} = j | X_n = h) P(X_n = h) =$$

$$= \frac{1}{2} \left(P(X_n = j - 1) + P(X_n = j + 1)\right)$$

$$(4.16)$$

Subtracting $P(X_n = j)$ from both members yields:

$$P(X_{n+1} = j) - P(X_n = j) = \frac{1}{2} \left(P(X_n = j - 1) + P(X_n = j + 1) - 2P(X_n = j)\right) \quad (4.17)$$

Equation has the form of a partial differential equation with time and space finite differences instead of derivatives, closely resembling the celebrated heat equation, which plays a central role in the study of diffusive processes like, for example, the flow of heat through a material:

$$\frac{\partial p}{\partial t} = D \frac{\partial^2 p}{\partial x^2} \qquad (4.18)$$

We stress that the analogy is not limited to one dimension, but is valid in arbitrary dimension.

Another similarity between the random walk and the heat equation is make evident by the following equalities, resulting from easy calculations:

$$E[X_n] = \sum_{i=0}^{n} E[\xi_i] = 0$$

$$(4.19)$$

$$Var(X_n) = \sum_{i=0}^{n} Var(\xi_i) = n$$

The mean distance covered by the walker scales with the square root of the number of steps, a typical property of diffusion processes.

The connection between a stochastic process and a partial differential equation is not a coincidence, but the first appearence of a general relationship between two apparently disconnected fields of Mathematics, which we will explore in detail in the next chapters.

4.2.3 The Asymptotic Behavior for $n \to +\infty$ and Recurrence

We observe that:

$$\left| \frac{1}{d} \sum_{\mu=1}^{d} \cos(k_\mu) \right| \leq 1, \quad \mathbf{k} \in [-\pi, \pi]^d \tag{4.20}$$

the left hand side taking the value 1 only if $\mathbf{k} = 0$. This implies that the behavior of the integral when $n \to +\infty$ is governed by a small region near $\mathbf{k} = 0$, where we can make the expansion:

$$\frac{1}{d} \sum_{\mu=1}^{d} \cos(k_\mu) = 1 - \frac{1}{2d} |\mathbf{k}|^2 + \cdots \tag{4.21}$$

yielding the approximation:

$$P\left(X_n = \mathbf{x}\right) \simeq \frac{1}{(2\pi)^d} \int_{\mathbb{R}^d} d\mathbf{k} \, \exp\left(i\mathbf{k} \cdot \mathbf{x} - \frac{n}{2d} |\mathbf{k}|^2 \right) \tag{4.22}$$

where we have extended the integration to the whole $\mathbb{R}^d$ since the exponential term guarantees that only vectors $\mathbf{k}$ near the origin play a significant role. This integral can be performed analytically, giving the result:

$$P\left(X_n = \mathbf{x}\right) \simeq \left(\frac{d}{2\pi n} \right)^{d/2} \exp\left(-2d \frac{|\mathbf{x}|^2}{4n} \right) \tag{4.23}$$

The above expression is very interesting. It shows that, for large time n, the law of the random walk becomes normal $N(0, n/d)$. Moreover, let's consider the number of times, say N, the walker comes back to visit the origin $\mathbf{x} = 0$. We write:

$$N = \sum_{n=1}^{+\infty} I_n \tag{4.24}$$

where $I_n = 1$ if $X_n = 0$ while $I_n = 0$ otherwise. We have:

$$E[N] = \sum_{n=1}^{+\infty} E[I_n] = \sum_{n=1}^{+\infty} P\,(X_n = 0) \tag{4.25}$$

We have found that:

$$P\,(X_n = 0) \sim \frac{1}{n^{d/2}} \tag{4.26}$$

This implies that, when $d = 1, 2$, $E[N] = +\infty$, while, when $d \geq 3$, $E[N] \leq \infty$. This is a very interesting feature: when $d = 1, 2$ we have an infinite expected number of returns to the origin, while this is not true in higher dimensions. Typically a random walk giving an infinite expected number of returns to the origin is called **recurrent**, while, when $E[N] < +\infty$, the random walk is called **transient**.

4.3 Recursive Markov Chains

The random walk on $\mathbb{Z}^d$ exemplifies a general procedure for constructing explicitly Markov chains. Let X_0 be a given random variable, and $\{U_m\}_{m \in \mathbb{N}}$ a sequence of independent and identically distributed uniform random variables in $(0, 1)$. Let moreover $h : E \times (0, 1) \to E$ be a function, until now arbitrary. We define:

$$X_{n+1} = h\,(X_n, U_{n+1}) \tag{4.27}$$

where U_{n+1} is assumed to be independent from X_n. Repeating the calculation of the previous paragraph it is easily concluded that $\{X_n\}_n$ is a Markov chain. Its transition matrix is readily obtained:

$$P(X_{n+1} = j | X_n = i) = \frac{P(h\,(X_n, U_{n+1}) = j, X_n = i)}{P(X_n = i)} = \tag{4.28}$$

$$= \frac{P(h\,(i, U_{n+1}) = j, X_n = i)}{P(X_n = i)} = P(h\,(i, U_{n+1}) = j) = \mathcal{P}_{i \to j}$$

This result is very useful since, as we will explain in the next chapter, having at our disposal a random number generator, the problem of simulating a Markov chain with initial law v and transition matrix $\mathcal{P}$ is solved sampling the initial state with probability v and iteratively applying $X_{n+1} = h\,(X_n, U_{n+1})$ where $h : E \times (0, 1) \to E$ is a function such that $P(h\,(i, U) = j) = \mathcal{P}_{i \to j}$ if U is uniform in $(0, 1)$.

Remark 4.1 From now on we will always assume that set E is finite. For the sake of simplicity, we will write $E = \{1, \ldots, N\}$. Probabilities on the set E, when useful, will be identified with row vectors $\mathbf{v} = (v_1, \ldots, v_N) \in \mathbb{R}^N$, $v_i \geq 0$, $\sum_{i=1}^N v_i = 1$.

4.4 Transition Matrix and Initial Law

We will now show how initial law and transition matrix give an exhaustive knowledge of the corresponding homogeneous Markov chain. We begin computing the law $v^{(1)}$ of X_1:

$$P(X_1 = k) = \sum_{h=1}^{N} P(X_0 = h) P(X_1 = k | X_0 = h) = \sum_{h=1}^{N} v_h \, \mathcal{P}_{h \to k} \qquad (4.29)$$

which can be written in matrix form recalling that laws can be represented through row vectors in $\mathbb{R}^N$:

$$\mathbf{v}^{(1)} = \mathbf{v} \, \mathcal{P} \qquad (4.30)$$

At the subsequent instant:

$$\begin{aligned}
P(X_2 = k) &= \sum_{l=1}^{N} P(X_1 = l) P(X_2 = k | X_1 = l) = \\
&= \sum_{l=1}^{N} P(X_1 = l) \, \mathcal{P}_{l \to k} = \sum_{l=1}^{N} \sum_{h=1}^{N} v_h \, \mathcal{P}_{h \to l} \, \mathcal{P}_{l \to k} = \\
&= \sum_{h=1}^{N} v_h \sum_{l=1}^{N} \mathcal{P}_{h \to l} \, \mathcal{P}_{l \to k}
\end{aligned} \qquad (4.31)$$

that is:

$$\mathbf{v}^{(2)} = \mathbf{v} \, \mathcal{P}^2 \qquad (4.32)$$

Iterating this reasoning we easily conclude that the law at instant n is obtained applying to the row vector representing the initial law the n-th power of the transition matrix:

$$\mathbf{v}^{(n)} = \mathbf{v} \, \mathcal{P}^n \qquad (4.33)$$

It is interesting to observe that, denoting with $\mathcal{P}^{(m)}_{i \to j}$ the matrix elements of the m-th power $\mathcal{P}^m$ of the transition matrix, one obtains the m-step transition probabilities.

$$\mathcal{P}^{(m)}_{i \to j} = P(X_{n+m} = j | X_n = i) \qquad (4.34)$$

This can be shown iterating the following calculation:

$$\begin{aligned}
P(X_{n+m} = j | X_n = i) &= \frac{P(X_{n+m} = j, X_n = i)}{P(X_n = i)} = \\
&= \sum_h \frac{P(X_{n+m} = j, X_{n+m-1} = h, X_n = i)}{P(X_n = i)} = \\
&= \sum_h \frac{P(X_{n+m} = j, X_{n+m-1} = h, X_n = i)}{P(X_{n+m-1} = h, X_n = i)} \frac{P(X_{n+m-1} = h, X_n = i)}{P(X_n = i)} = \\
&= \sum_h P(X_{n+m} = j | X_{n+m-1} = h, X_n = i) P(X_{n+m-1} = h | X_n = i) = \\
&= \sum_h \mathcal{P}_{h \to j} P(X_{n+m-1} = h | X_n = i)
\end{aligned} \qquad (4.35)$$

We eventually compute the joint laws ot the process in terms of the initial law v and of the transition matrix $\mathcal{P}$, that is $P\left(X_{n_1} = i_1, \ldots, X_{n_k} = i_k\right), 0 \leq n_1 < \cdots < n_k$:

$$P\left(X_{n_1} = i_1, \ldots, X_{n_k} = i_k\right) =$$
$$= P\left(X_{n_1} = i_1, \ldots, X_{n_{k-1}} = i_{k-1}\right) P\left(X_{n_k} = i_k | X_{n_1} = i_1, \ldots, X_{n_{k-1}} = i_{k-1}\right) =$$
$$= P\left(X_{n_1} = i_1, \ldots, X_{n_{k-1}} = i_{k-1}\right) \mathcal{P}^{(n_k - n_{k-1})}_{i_{k-1} \to i_k} = \cdots = \qquad (4.36)$$
$$= P\left(X_{n_1} = i_1\right) \mathcal{P}^{(n_2 - n_1)}_{i_1 \to i_2} \cdots \mathcal{P}^{(n_k - n_{k-1})}_{i_{k-1} \to i_k} =$$
$$= \sum_j v_j \mathcal{P}^{(n_1)}_{j \to i_i} \mathcal{P}^{(n_2 - n_1)}_{i_1 \to i_2} \cdots \mathcal{P}^{(n_k - n_{k-1})}_{i_{k-1} \to i_k}$$

Remark 4.2 Rewriting the 2- and 3- times joint laws in the form:

$$P\left(X_{n_1} = i_1, X_{n_3} = i_3\right) = P\left(X_{n_1} = i_1\right) \mathcal{P}^{(n_3 - n_1)}_{i_1 \to i_3} \qquad (4.37)$$

$$P\left(X_{n_1} = i_1, X_{n_2} = i_2, X_{n_3} = i_3\right) = P\left(X_{n_1} = i_1\right) \mathcal{P}^{(n_2 - n_1)}_{i_1 \to i_2} \mathcal{P}^{(n_3 - n_2)}_{i_2 \to 3} \qquad (4.38)$$

the relation:

$$P\left(X_{n_1} = i_1, X_{n_3} = i_3\right) = \sum_{i_2=1}^{N} P\left(X_{n_1} = i_1, X_{n_2} = i_2, X_{n_3} = i_3\right) \qquad (4.39)$$

is equivalent to the followig **Chapman-Kolmogorov equation** for the m-step transition probability:

$$\mathcal{P}^{(n_3 - n_1)}_{i_1 \to i_3} = \sum_{i_2=1}^{N} \mathcal{P}^{(n_2 - n_1)}_{i_1 \to i_2} \mathcal{P}^{(n_3 - n_2)}_{i_2 \to 3}, \quad 0 \le n_1 < n_2 < n_3 \qquad (4.40)$$

The property (4.40), quite natural in the present context since it is known that the m-step transition probability is obtained computing the m-th power of the transition matrix, will turn out of great importance when dealing with continous-time Markov processes taking values in $\mathbb{R}^d$.

4.5 Invariant Laws

Given a probability distribution π on E, which as seen before can be represented with a row vector $\pi = (\pi_1, \ldots, \pi_N) \in \mathbb{R}^N$, and a homogeneous Markov chain with transition matrix $\mathcal{P} = \{\mathcal{P}_{i \to j}\}_{i,j}$ and initial law v, we will say that π is **invariant** provided that:

$$\pi = \pi \mathcal{P}, \quad i.e. \quad \pi_k = \sum_{h \in E} \pi_h \mathcal{P}_{h \to k} \qquad (4.41)$$

We stress that if the initial law v is invariant, X_n has law v for all n: all the X_n have the same law, giving rise to a **stationary Markov process**.

We now prove an important result:

Theorem 4.1 (Markov-Kakutani) *Any transition matrix $\mathcal{P}$ admits has at least an invariant law.*

Proof We first observe that there is a one-to-one correspondence between probabilities on E ad points in the following set, the simplex:

$$S = \left\{ \mathbf{x} \in \mathbb{R}^N : 0 \le x_i \le 1, \sum_{i=1}^{N} x_i = 1 \right\} \tag{4.42}$$

S is a closed and limited set in $\mathbb{R}^N$, and therefore compact: by virtue of Bolzano-Weierstrass theorem of classical Mathematical Analysis [1], any sequence in S has a convergent subsequence. Given a generic point $\mathbf{x} \in S$, consider the sequence:

$$\mathbf{x}_n = \frac{1}{n} \sum_{k=0}^{n-1} \mathbf{x}\,\mathcal{P}^k \tag{4.43}$$

Obviously $\mathbf{x}_n$ has non-negative components. Moreover, $\mathbf{x}_n \in S$ as the following simple calculation shows:

$$\sum_i x_{n,i} = \frac{1}{n} \sum_{k=0}^{n-1} \sum_h \sum_i x_h \mathcal{P}^{(k)}_{h \to i} = \frac{1}{n} \sum_{k=0}^{n-1} \sum_h x_h = 1 \tag{4.44}$$

where we have taken into account the fact that $\mathcal{P}^{(k)}_{h \to i}$ is the probability of moving from h to i in k steps, and therefore $\sum_i \mathcal{P}^{(k)}_{h \to i} = 1$.

Since $\{\mathbf{x}_n\}_n \subset S$ it has a subsequence: $\{\mathbf{x}_{n_k}\}_{n_k}$ converging to a point $\pi \in S$. We observe that:

$$\mathbf{x}_{n_k} - \mathbf{x}_{n_k}\mathcal{P} = \frac{1}{n_k} \left(\sum_{h=0}^{n_k-1} \mathbf{x}\,\mathcal{P}^h - \sum_{h=0}^{n_k-1} \mathbf{x}\,\mathcal{P}^{h+1} \right) = \frac{1}{n_k} \left(\mathbf{x} - \mathbf{x}\,\mathcal{P}^{n_k} \right) \tag{4.45}$$

and since the quantity $\mathbf{x} - \mathbf{x}\,\mathcal{P}^{n_k}$ is limited by construction:

$$\pi - \pi\,\mathcal{P} = \lim_{k \to +\infty} \left(\mathbf{x}_{n_k} - \mathbf{x}_{n_k}\mathcal{P} \right) = \lim_{k \to +\infty} \frac{1}{n_k} \left(\mathbf{x} - \mathbf{x}\,\mathcal{P}^{n_k} \right) = 0 \tag{4.46}$$

which completes the proof.

We observe that the proof of Markov-Kakutani's theorem is constructive: any sequence $\mathbf{x}_n$, once thinned out, converges to an invariant law, which may be not unique. Since the point $\mathbf{x} = \mathbf{x}_0$ is completely arbitrary, it can be chosen $x_h = \delta_{h,i}$, corresponding to a probability distribution concentrated at the point i. Were that the case:

$$x_{n,j} = \frac{1}{n} \sum_{k=0}^{n-1} \mathcal{P}_{i \to j}^{(k)} \tag{4.47}$$

If $\{X_n\}$ is the Markov chain with initial law $\mathbf{x}$, concentrated at the point i with transition matrix $\mathcal{P}$, we know that:

$$\mathcal{P}_{i \to j}^{(k)} = P(X_k = j) \tag{4.48}$$

Therefore:

$$x_{n,j} = \frac{1}{n} \sum_{k=0}^{n-1} \mathcal{P}_{i \to j}^{(k)} = \frac{1}{n} \sum_{k=0}^{n-1} P(X_k = j) = E\left[\frac{1}{n} \sum_{k=0}^{n-1} 1_{\{X_k=j\}} \right] \tag{4.49}$$

and $x_{n,j}$ coincides with the expectation of the random variable:

$$\frac{1}{n} \sum_{k=0}^{n-1} 1_{\{X_k=j\}} \tag{4.50}$$

representing the fraction of time the process has spent in the state j before the n-th time step. Remarkably, for large n the expectation of such random variable approximates the j-th component of one invariant law.

To compute the invariant law(s), the following problem must be solved:

$$\pi_j = \sum_{i=1}^{n} \pi_i \, \mathcal{P}_{i \to j} \tag{4.51}$$

To this purpose, the following interesting result holds, which represent a sufficient condition for a law π to be invariant:

Theorem 4.2 *If a law π satisfies the* **detailed balance equation***:*

$$\pi_i \, \mathcal{P}_{i \to j} = \pi_j \, \mathcal{P}_{j \to i}, \quad \forall i, j \in E \tag{4.52}$$

then it is invariant.

Proof The proof is simple:

$$\sum_{i=1}^{n} \pi_i \, \mathcal{P}_{i \to j} = \sum_{i=1}^{n} \pi_j \, \mathcal{P}_{j \to i} = \pi_j \qquad (4.53)$$

A transition matrix may have, in general, infinite invariant laws: in fact, as a simple calculation shows, if π and π' are distinct stationary laws for $\mathcal{P}$, any convex linear combination of π and π' is still a stationary law for $\mathcal{P}$.

It is therefore very interesting to investigate conditions for the uniqueness of the invariant law. To this purpose, we introduce the following definitions:

Definition 4.3 Let $\mathcal{P} = \{\mathcal{P}_{i \to j}\}_{i,j}$ be the transition matrix of a homogeneous Markov chain.

1. a state $j \in E$ is accessible from $i \in E$, denoted $i \to j$ if there exists an integer $m \geq 0$ such that $\mathcal{P}_{i \to j}^{(m)} > 0$. Conversely, if $\mathcal{P}_{i \to j}^{(m)} = 0$ for all $m \geq 0$, then the state j is not accessible from i, denoted $i \nrightarrow j$.
2. the states $j, i \in E$ communicate if j is accessible from i and viceversa, denoted $i \leftrightarrows j$.

Remark 4.3 It is useful to include in the exposition the zeroth power $\mathcal{P}_{i \to j}^{(0)} = \delta_{i,j}$, which is actually a trivial zero steps transition matrix, to deal with communications among states.

The communication relation $\leftrightarrows$ satisfies the following conditions:

1. reflexivity: $\forall i \in E, i \leftrightarrows i$ since $\mathcal{P}_{i \to i}^{(0)} = 1$
2. symmetry: $\forall i, j \in E, i \leftrightarrows j$ if and only if $j \leftrightarrows i$
3. transitivity: $\forall i, j, k \in E$, if $i \leftrightarrows j$ and $j \leftrightarrows k$ then $i \leftrightarrows k$

To prove the transitivity condition, let us observe that if $i \leftrightarrows j$ and $j \leftrightarrows k$ there exists two integer numbers m, n such that $\mathcal{P}_{i \to j}^{(m)} > 0$ and $\mathcal{P}_{j \to k}^{(n)} > 0$. Therefore:

$$\mathcal{P}_{i \to k}^{(m+n)} = \sum_{l} \mathcal{P}_{i \to l}^{(m)} \mathcal{P}_{l \to k}^{(n)} \geq \mathcal{P}_{i \to j}^{(m)} \mathcal{P}_{j \to k}^{(n)} > 0 \qquad (4.54)$$

The above conditions imply that communication is an equivalence relation. The state space E can be uniquely decomposed into mutually disjoint subsets $\{E_i\}_i$ whose union equals E, the equivalence classes of the homogeneous Markov chain.

Definition 4.4 Let $\mathcal{P} = \{\mathcal{P}_{i \to j}\}_{i,j}$ be the transition matrix of a homogeneous Markov chain.

1. $\mathcal{P} = \{\mathcal{P}_{i \to j}\}$ is **irreducibile** if all states in E communicate with each other.
2. $\mathcal{P} = \{\mathcal{P}_{i \to j}\}$ is **regular** if there exists a number $m > 0$ such that $\mathcal{P}_{i \to j}^{(m)} > 0$ for all $i, j \in E$.

A regular trasition matrix is always irreducible, but the converse is not true in general. Nevertheless, the following result holds:

Lemma 4.1 *If a transition matrix is irreducible, and there exists $h \in E$ such that $\mathcal{P}_{h \to h} > 0$, it is regular.*

Proof If for all $i, j \in E$ there exists $m = m(i, j) \geq 0$ such that $\mathcal{P}^{(m)}_{i \to j} > 0$, chosen $s = \max_{i,j \in E} m(i, j)$ we have $\mathcal{P}^{(2s)}_{l \to k} > 0$ for all $l, k \in E$, as the following inequality makes clear:

$$\mathcal{P}^{(2s)}_{l \to k} \geq \mathcal{P}^{(n(l,h))}_{l \to h} \mathcal{P}_{h \to h} \cdots \mathcal{P}_{h \to h} \mathcal{P}^{(n(h,k))}_{h \to k} > 0 \qquad (4.55)$$

in which the term $\mathcal{P}_{h \to h}$ appears $2s - n(l, h) - n(h, k)$ times.

Remark 4.4 At a first glance, it might seem very difficult to verify whether a chain is irreducible or not, but there exist observations that can considerably simplify the calculations involved. Chosen two states $i, j, i \neq j$, if the chain is irreducible there exists $m > 0$, in general depending on the couple (i, j) of interest, such that $\mathcal{P}^{(m)}_{i \to j} > 0$; since the transition matrix has non-negative elements, this corresponds to the existence of at least one $(m - 1)$-tuple of states $k_1, \ldots, k_{m-1}$ such that:

$$0 < \mathcal{P}_{i \to k_1} \mathcal{P}_{k_1 \to k_2} \cdots \mathcal{P}_{k_{m-1} \to j} \leq \mathcal{P}^{(m)}_{i \to j} \qquad (4.56)$$

Intuitively, it is necessary to move from i to j passing through points in E making steps with non-zero transition probability. At least for Markov chains with small state space, this can be checked representing the Markov chain as a directed graph $\Gamma = (E, V)$ with links $(i, j) \in V \subseteq E \times E$ connecting states $i, j \in E$ for which $\mathcal{P}_{i \to j} > 0$.

This pictorial representation permits to quickly verify whether a state j is accessible from another state i. In the following example:

$$\mathcal{P} = \begin{pmatrix} 0 & 1 & 0 & 0 & 0 & 0 \\ 0 & \mathcal{P}_{2 \to 2} & 0 & 0 & \mathcal{P}_{2 \to 5} & 0 \\ 0 & 0 & 0 & 0 & 1 & 0 \\ 0 & \mathcal{P}_{4 \to 2} & 0 & 0 & \mathcal{P}_{4 \to 5} & 0 \\ \mathcal{P}_{5 \to 1} & \mathcal{P}_{5 \to 2} & 0 & 0 & 0 & 0 \\ 0 & 0 & 0 & 0 & 0 & 1 \end{pmatrix}$$

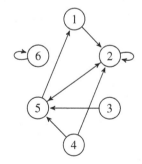

the simple observation of the graph Γ shows that $\mathcal{P}$ is not irreducible. The equivalence classes into which E is split by the communication equivalence relation are $\{1, 2, 5\}, \{3\}, \{4\}, \{6\}$.

We are going now to present the most important result, involving regular transition matrices, which is the Markov theorem, which states that the Markov chain thermalizes, in the sense that, for every initial law, the law of X_n, for $n \to +\infty$, converges to the unique *equilibrium* invariant law.

We need first the following preliminary version of Markov theorem:

Theorem 4.3 *If a transition matrix $\mathcal{P}$ has all strictly positive entries, it admits a unique invariant law π^* and, for all initial laws v:*

$$\pi_j^* = \lim_{n \to +\infty} \left(v \, \mathcal{P}^n \right)_j \qquad (4.57)$$

Proof By virtue of Markov-Kakutani theorem, $\mathcal{P}$ admits an invariant law π^*. By definition, it is a fixed point of the map:

$$\mathcal{C} : S \to S, \quad v \mapsto \mathcal{C}(v) = v \mathcal{P} \qquad (4.58)$$

where, as before:

$$S = \left\{ \mathbf{x} \in \mathbb{R}^N : 0 \le x_i \le 1, \ \sum_{i=1}^N x_i = 1 \right\} \qquad (4.59)$$

It will now be shown that $\mathcal{C}$ is a strict contraction on S relative to the following distance:

$$d(v, w) = \frac{1}{2} \sum_{i \in E} |v_i - w_i| \qquad (4.60)$$

First, since all the entries of $\mathcal{P}$ are strictly positive, there exists some number ε such that $\varepsilon < \frac{1}{N}$ and $\mathcal{P}_{i \to j} \ge \varepsilon \ \forall i, j \in E$. It is simple to show that $\mathcal{Q}_{i \to j} = \frac{\mathcal{P}_{i \to j} - \varepsilon}{1 - N\varepsilon}$ is a transition matrix on E. The distance $d(\mathcal{C}(v), \mathcal{C}(w))$ between the images of two generic laws $v, w \in S$ through the map $\mathcal{C}$ can be expressed as:

$$\frac{1}{2} \sum_j \left| (v \, \mathcal{P})_j - (w \, \mathcal{P})_j \right| = \frac{1}{2} \sum_j \left| \sum_i (v_i - w_i)(1 - N\varepsilon) \mathcal{Q}_{i \to j} \right| \qquad (4.61)$$

and since:

$$\left| \sum_i (v_i - w_i)(1 - N\varepsilon) \mathcal{Q}_{i \to j} \right| \le (1 - N\varepsilon) \sum_i |v_i - w_i| \, \mathcal{Q}_{i \to j} \qquad (4.62)$$

and $\mathcal{Q}$ is a transition matrix, the distance Eq. (4.61) is bounded by:

$$d\left(\mathcal{C}(v), \mathcal{C}(w)\right) \le (1 - N\varepsilon) \sum_i |v_i - w_i| \sum_j \mathcal{Q}_{i\to j} =$$
$$= (1 - N\varepsilon) \sum_i |v_i - w_i| = (1 - N\varepsilon)\, d(v, w) \tag{4.63}$$

Equation (4.63) ensures that $\mathcal{C}$ is a contraction. Hence the uniqueness of the invariant law π^* and its expression Eq. (4.64) follow from Banach's fixed point theorem [1].

The Theorem 4.3 has been proved under the strict requirement that all the entries of $\mathcal{P}$ are positive. Such condition can be relaxed, leading to the following fundamental result:

Theorem 4.4 (Markov) *If a transition matrix $\mathcal{P}$ is regular, it admits a unique invariant law π^* and, for all initial laws v:*

$$\pi_j^* = \lim_{n\to+\infty} \left(v\,\mathcal{P}^n\right)_j \tag{4.64}$$

Proof The regularity condition implies the existence of an integer $m > 0$ such that $\mathcal{P}^m$ has all entries strictly positive. Hence, by Banach's fixed point theorem there exists a unique law π^* such that:

$$\pi^* = \pi^*\mathcal{P}^m \qquad \lim_{n\to\infty} v\,\mathcal{P}^{mn} = \pi^* \quad \forall v \in S \tag{4.65}$$

the second of eqs. (4.65) still holds if we change $v \to v\,\mathcal{P}^k$ with $k < m - 1$. Now, if a sequence $\{v_n\}_n$ has the property that all the subsequences $\{v_{k+mn}\}_n$, with $k < m - 1$, converge to the same limit v^*, then it converges to v^*. Therefore:

$$\lim_{n\to\infty} v\,\mathcal{P}^n = \pi^* \quad \forall v \in \mathcal{S}_E \tag{4.66}$$

π^* is also invariant since:

$$\pi^*\mathcal{P} = \left(\lim_{n\to\infty} v\,\mathcal{P}^n\right)\mathcal{P} = \lim_{n\to\infty} v\,\mathcal{P}^{n+1} = \pi^* \tag{4.67}$$

by the continuity of the matrix product.

Markov's Theorem 4.4 ensures that if a homogeneous Markov chain with regular transition matrix starts at a generic initial law v, it converges exponentially fast to a unique law π^*.

4.6 Metropolis Theorem

Consider a given probability distribution π on E. We are going to somehow reverse the point of view: we wonder whether there exists a transition matrix $\mathcal{P}$ such that, for all initial laws ν

$$\pi_j = \lim_{n \to +\infty} \left(\mathbf{v}\, \mathcal{P}^n \right)_j \tag{4.68}$$

Were that the case, we could construct a Markov chain $\{X_n\}_n$ with law converging, as n tends to infinity, to π in the sense precised by (4.68).

As the reader might have guessed, this possibility has deep implications in the field of simulations.

To this purpose, it turns out to be necessary to assume that $\pi_j > 0$ for all the states $j \in E$ and that π is not the uniform distribution.

Let now $\mathcal{T} = \left\{ \mathcal{T}_{i \to j} \right\}$ be a symmetric and irreducible transition matrix, $\mathcal{T}_{i \to j} = \mathcal{T}_{j \to i}$, subject to no other restrictions, and define:

$$\mathcal{P}_{i \to j} = \begin{cases} \mathcal{T}_{i \to j}, & i \neq j, \ \pi_j \geq \pi_i \\ \mathcal{T}_{i \to j} \frac{\pi_j}{\pi_i}, & i \neq j, \ \pi_j < \pi_i \\ 1 - \sum_{j \neq i} \mathcal{P}_{i \to j}, & i = j \end{cases} \tag{4.69}$$

We have the following [2]:

Theorem 4.5 *If $\pi_j > 0$ for all the states $j \in E$ and π is not uniform, for all the initial laws ν the Markov chain $\{X_n\}_n$ with initial law ν and transition matrix $\mathcal{P}$ is regular, and has π as unique invariant distribution. Consequently:*

$$\pi_j = \lim_{n \to +\infty} \left(\mathbf{v}\, \mathcal{P}^n \right)_j = \lim_{n \to +\infty} P(X_n = j) \tag{4.70}$$

Proof We start showing that the detailed balance condition is satisfied. We choose two states (i, j) such that, without loss of generality, $\pi_j \leq \pi_i$, $\mathcal{P}_{i \to j} = \mathcal{T}_{i \to j} \frac{\pi_j}{\pi_i}$ whereas $\mathcal{P}_{j \to i} = \mathcal{T}_{j \to i}$ and thus:

$$\pi_i \mathcal{P}_{i \to j} = \pi_i \mathcal{T}_{i \to j} \frac{\pi_j}{\pi_i} = \mathcal{T}_{i \to j} \pi_j = \pi_j \mathcal{P}_{j \to i} \tag{4.71}$$

where the hypothesis that $\mathcal{T}$ is symmetric has been used. As a consequence of (4.71) π is invariant.

It remains to show that the Markov chain (4.69) is regular; first, we show that it is irreducible. In fact, if $i \neq j$ and $\mathcal{T}_{i \to j} > 0$, then, by construction $\mathcal{P}_{i \to j} > 0$; this means that, if $\mathcal{T}$ makes two states communicate, the same does $\mathcal{P}$, and this guarantees that $\mathcal{P}$ is irreducible.

To prove that (4.69) is regular, by virtue of lemma (4.1) it is sufficient to show that there exists $i_0 \in E$ such that $\mathcal{P}_{i_0 \to i_0} > 0$. Since π is not uniform, there exists a proper subset $M \subset E$, $M \neq E$ of E on which π takes maximum value; due to the irreducibility of $\mathcal{T}$ the chain can move outside M, and therefore there exist $i_0 \in M$ and $j_0 \in M^c$ such that $\mathcal{T}_{i_0 \to j_0} > 0$ and, by construction, $\pi_{i_0} > \pi_{j_0}$. Moreover, $\mathcal{P}_{i \to j} \leq \mathcal{T}_{i \to j}$ if $i \neq j$. These intermediate results imply:

$$
\begin{aligned}
\mathcal{P}_{i_0 \to i_0} &= 1 - \sum_{j \neq i_0} \mathcal{P}_{i_0 \to j} = 1 - \sum_{j \neq i_0, j_0} \mathcal{P}_{i_0 \to j} - \mathcal{P}_{i_0 \to j_0} \geq \\
&\geq 1 - \sum_{j \neq i_0, j_0} \mathcal{T}_{i_0 \to j} - \mathcal{T}_{i_0 \to j_0} \frac{\pi_{j_0}}{\pi_{i_0}} = \\
&= 1 - \sum_{j \neq i_0} \mathcal{T}_{i_0 \to j} + \mathcal{T}_{i_0 \to j_0}\left(1 - \frac{\pi_{j_0}}{\pi_{i_0}}\right) = \mathcal{T}_{i_0 \to i_0} + \mathcal{T}_{i_0 \to j_0}\left(1 - \frac{\pi_{j_0}}{\pi_{i_0}}\right) \geq \\
&\geq \mathcal{T}_{i_0 \to j_0}\left(1 - \frac{\pi_{j_0}}{\pi_{i_0}}\right) > 0
\end{aligned}
\tag{4.72}
$$

that is, the chain is regular by virtue of Lemma (4.1) and admits a unique stationary law by virtue of Markov's theorem.

Metropolis' theorem is widely used in Physics, where it provides a technique for simulating random variables with law given by:

$$
\pi_i = \frac{e^{-\beta \mathcal{H}(i)}}{Z(\beta)}, \quad \mathcal{H} : E \to \mathbb{R}, \quad Z(\beta) = \sum_{i \in E} e^{-\beta \mathcal{H}(i)}
\tag{4.73}
$$

E being the configuration space of the classical system under study and $\mathcal{H}$ its Hamiltonian. Notice that the knowledge of $Z(\beta)$ (resulting from an integration procedure which, for large systems of interacting particles, cannot be preformed neither analytically nor numerically) is not necessary for applying (4.69). We will discuss the application in some detail in the next chapter.

Usually (4.69) is written, for $i \neq j$, in the form:

$$
\mathcal{P}_{i \to j} = \mathcal{T}_{i \to j} A_{i \to j}, \quad A_{i \to j} = \min\left(1, \frac{\pi_j}{\pi_i}\right)
\tag{4.74}
$$

$\mathcal{T}_{i \to j}$ is a *trial move* that is *accepted or refused* depending on the outcome of a *Metropolis test* controlled by the term $\min\left(1, \frac{\pi_j}{\pi_i}\right)$.

We remind the reader that the hypothesis $\mathcal{T}_{i \to j} = \mathcal{T}_{j \to i}$ was framed in the proof of Metropolis' theorem. This hypothesis can be removed, provided that $\mathcal{T}_{j \to i} > 0$ whenever $\mathcal{T}_{i \to j} > 0$; in such situation, Metropolis' theorem still holds for the Markov chain:

$$
\mathcal{P}_{i \to j} = \mathcal{T}_{i \to j} A_{i \to j}, \quad A_{i \to j} = \min\left(1, \frac{\pi_j \mathcal{T}_{j \to i}}{\pi_i \mathcal{T}_{i \to j}}\right) \quad i \neq j
\tag{4.75}
$$

where it is meant that $\mathcal{P}_{i \to j} = 0$ if $\mathcal{T}_{i \to j} = 0$, whereas $\mathcal{P}_{i \to i}$ is defined as in (4.69).

4.7 Further Readings

The topic of Markov chains and their application is extremely vast, and dealt with with many textbooks. Readers interested in deepening their knowledge of the mathematical background can see, for example, [2]. For applications in simulation of physical systems, we refer to the bibliography of the following chapter.

Problems

4.1 Ehrenfest model
Suppose there are N particles in two containers. At time $n = 0$, all the particles are in the first container and, subsequently, they can diffuse changing container. Let X_n the number of particles in the first container at time instant n. Assume that the time evolution is markovian, driven by a transition matrix of the form:

$$\mathcal{P}_{i \to j} = \begin{cases} q_i = i/N & \text{if } j = i - 1 \\ p_i = (N - i)/N & \text{if } j = i + 1 \\ 0, & otherwise \end{cases}$$

Find the invariant law of the Markov chain.

4.2 Random walk on a triangle
Study a random walk on the vertices of an equilater triangle, with transition matrix:

$$\mathcal{P} = \begin{pmatrix} 0 & p & 1-p \\ 1-p & 0 & p \\ p & 1-p & 0 \end{pmatrix}$$

4.3 Galton-Watson process
The Galton-Watson process is a branching stochastic process describing the extinction of family names. Let the time n enumerate the generations and Z_n be the number of individuals with a given family name at generation n.

We model the situation as follows: if $\xi_m^{(n)}$ denote the number of descendants of the i-th individual at generation n, we will have:

$$Z_{n+1} = \sum_{i=1}^{Z_n} \xi_i^{(n)} \tag{4.76}$$

We have:

$$P(Z_{n+1} = m | Z_n = k) = \begin{cases} P\left(\sum_{i=1}^{k} \xi_i^{(n)} = m\right), & k > 0 \\ \delta_{m,0}, & k = 0 \end{cases} \tag{4.77}$$

Assuming $\{\xi_i^{(n)}\}_i$ independent and identically distributed with density p, assumed to be independent from n, we will have: $\mathcal{P}_{k\to m} = (p * \cdots * p)$, where:

$$(p * p)_i = \sum_{j\in E} p_j p_{i-j} \tag{4.78}$$

Study the Markov chain Z_n with state space $E = \{0, 1, 2, 3, \ldots, \}$ and compute $E[Z_{n+1}|Z_n]$ and $E[Z_{n+1}^2|Z_n]$, as well as the extinction probability starting from a single individual, defined as

$$P\left(\{\omega \in \Omega \mid \lim_{n\to+\infty} Z_n(\omega) = 0, \; Z_0(\omega) = 1\}\right) \tag{4.79}$$

In particular consider the Lotka probability density:

$$p_k = \begin{cases} 0.4825, & k = 0 \\ 0.2126\,(0.5893)^k, & k \neq 0 \end{cases} \tag{4.80}$$

4.4 Gambler's Ruin problem

Consider a gambler (player A) whose initial fortune is a coins, who plays against another player, say B, having an initial fortune of b coins: $a + b = N$. The rules of the game are as follows: at each time step player A either gives one coin to player b with probability p or he/she receives a coin from B with probability $q = 1 - p$. If, at a time instant, player A possesses N coins (win) or he/she runs out of coins (ruin), the game finishes. Write down a transition matrix describing the game, compute the probability that player A wins.

4.5 An explicit calculation

Consider the following transition matrix for a Markov chain with two accessible states:

$$\mathcal{P} = \begin{pmatrix} 1-a & a \\ b & 1-b \end{pmatrix}, \quad 0 < a, b < 1 \tag{4.81}$$

Show that:

$$\mathcal{P}^n = \frac{1}{a+b}\left\{\begin{pmatrix} b & a \\ b & a \end{pmatrix} + (1-a-b)^n \begin{pmatrix} a & -a \\ -b & b \end{pmatrix}\right\} \tag{4.82}$$

Evaluate explicitly $\lim_{n\to+\infty} \mathcal{P}^n$.

4.6 A monkey inside a labyrinth

A monkey is inside a labyrinth containing 9 cells. Assuming that the labyrinth is a square, at any time instant, the monkey changes cell with a probability proportional to the number of nearest neighboring cells. Write down the transition matrix and find the invariant law.

4.7 Monkey run

Suppose now that the monkey is allowed to escape from cell 9 of the labyrinth. Estimate the time that will take for the monkey to leave the labyrinth.

References

1. Rudin, W.: Real Complex Analysis. McGraw-Hill (1987)
2. Baldi, P. (1998). *Calcolo Delle Probabilitá e Statistica*. Milano: McGraw-Hill.

Chapter 5
Sampling of Random Variables and Simulation

Abstract In this chapter we introduce the art of sampling of random variables. Sampling a random variable X, for example real valued, means using a random number generator to generate n real numbers $(x_1, \ldots, x_n)$, realizations of a sample $(X_1, \ldots, X_n)$ of independent and identically distributed random variables sharing the same law as X. The ability of sampling is crucial to deal with integrals in high dimensions, appearing in quantum mechanics and in statistical physics, and gives the possibility to simulate physical systems. We first introduce simple tools to sample random variables, that can be used only in quite special situations. In the last part of the chapter, we will introduce and discuss a very general sampling technique, relying on the Metropolis theorem.

Keywords Sampling of random variables · Monte carlo integration
Random number generators · Metropolis algorithm · Simulation

5.1 Introduction

In our study of mathematical statistics, we have learnt to use data to infer the values of some parameters specifying the law of a random variable modeling the outcomes of the considered experiment. A set of data:

$$(\mathbf{x}_1, \ldots, \mathbf{x}_n) \tag{5.1}$$

are interpreted as realizations of a sample $(X_1, \ldots, X_n)$, that is:

$$(\mathbf{x}_1, \ldots, \mathbf{x}_n) = (X_1(\omega), \ldots, X_n(\omega)) \tag{5.2}$$

for a particular ω in some abstract probability space $(\Omega, \mathcal{F}, P)$ where the random variables are defined. This interpretation allows then to define estimators and confidence intervals, and to test some hypothesis regarding the outcome of the experiment.

In this chapter we will take the opposite point of view: given the law of a random variable X, or, equivalently, its probability density $p(\mathbf{x})$ (if it exists), is it possible to *generate* possible realizations of X? This is a central topic in the realm of simulations, and is usually called the **sampling** of random variables, or, equivalently, the **sampling** of probability densities.

Remark 5.1 In what follows we will use interchangeably the expressions sampling of a random variable, sampling of a law, and sampling of a probability density.

5.2 Monte Carlo Integration

One very important application of the sampling of probability densities is **Monte Carlo integration**. It is an extremely useful tool to evaluate integrals arising, for example, from statistical physics and quantum mechanics. It becomes quite the unique way to face multi-dimensional integrals, when typical strategies of numerical quadrature would require a huge number of operations, beyond the possibility of any computer.

Remark 5.2 Suppose to evaluate an integral in $d = 300$ dimensions, arising, for example, from classical statistical mechanics, where equilibrium physical properties of a system of 100 classical particles are expressed as integrals over the configuration space. A numerical quadrature scheme unavoidably requires a grid: if one chooses 10 points per dimension, then it is necessary to evaluate a function on 10^{300} points, which is enormous. The rule of thumb is that numerical quadrature methods hardly serve purpose when $d \geq 10$

To state the problem, let's consider an integral of the very general form:

$$I = \int_D d\mathbf{x}\, f(\mathbf{x})\, p(\mathbf{x}) \tag{5.3}$$

where $D \subset \mathbb{R}^d$, f is any (measurable) function and p is a probability density on D:

$$p(\mathbf{x}) \geq 0, \quad \int_D d\mathbf{x}\, p(\mathbf{x}) = 1 \tag{5.4}$$

The key observation is that, if X is a random variable having $p(\mathbf{x})$ as its probability density, the following equalities holds:

$$I = E\,[f(X)] \tag{5.5}$$

$$\int_D d\mathbf{x}\, (f(\mathbf{x}) - I)^2\, p(\mathbf{x}) = Var\,(f(X)) \tag{5.6}$$

The law of large numbers and the central limit theorem, provided that $E\,[f(X)]$ and $Var\,(f(X))$ are finite, guarantee that, if $\{X_i\}_{i=1}^{+\infty}$ is a sequence of **independent and identically distributed** random variables with density $p(\mathbf{x})$, we have:

$$I = \lim_{n \to +\infty} \frac{1}{n} \sum_{i=1}^{n} f(X_i) \tag{5.7}$$

and:

$$\frac{\frac{1}{n} \sum_{i=1}^{n} f(X_i) - I}{\sqrt{\frac{Var(f(X))}{n}}} \tag{5.8}$$

converges in distribution to a standard normal random variable $N(0, 1)$. Thus, if we are able to sample $p(\mathbf{x})$, i.e., in practice, to generate n points in D:

$$(\mathbf{x}_1, \ldots, \mathbf{x}_n) \tag{5.9}$$

realizations of $\{X_i\}_{i=1}^{n}$:

$$(\mathbf{x}_1, \ldots, \mathbf{x}_n) = (X_1(\omega), \ldots, X_n(\omega)) \tag{5.10}$$

for a ω in some abstract probability space $(\Omega, \mathcal{F}, P)$ where the random variables are defined, then we can evaluate:

$$I \simeq \frac{1}{n} \sum_{i=1}^{n} f(\mathbf{x}_i) \tag{5.11}$$

and use mathematical statistics to estimate confidence intervals for the exact value of the integral.

Precisely in the same way, finite or infinite summations:

$$I = \sum_{\mathbf{x}} f(\mathbf{x}) \, p(\mathbf{x}) \tag{5.12}$$

can be dealt with, whenever $p(\mathbf{x})$ is a discrete probability density.

To summarize, the problem of evaluating an integral is transferred into the problem of building up a (possibly large) number of points $(\mathbf{x}_1, \ldots, \mathbf{x}_n)$ starting from the knowledge of a probability density $p(\mathbf{x})$. Such problem is in general highly not trivial. The first necessary ingredient is an algorithm for the basic *generation of random numbers*.

5.3 Random Number Generators

Our starting point is the existence of the **random number generators**, which are algorithms able to sample a sequence of **independent uniform in** $(0, 1)$ random variables. The output of such an algorithm is a sequence:

$$(u_1, \ldots, u_n) \tag{5.13}$$

$0 < u_i < 1$, realizations of n **independent uniform in** $(0, 1)$ random variables:

$$(U_1, \ldots, U_n) \tag{5.14}$$

Remark 5.3 Sometimes, the generated random numbers lie in $[0, 1)$ or in $(0, 1]$, depending on the particular algorithm. It is important to be aware of this when applying functions to the random numbers, as we will see below.

We are not going now to enter the details of the theory of random number generation, requiring complex notions of numbers theory beyond the scope of this book. We simply mention the simplest algorithm, the **linear congruential generator** (LCG), introduced by D.H. Lehmer in 1949, which builds up the sequence $(u_1, \ldots, u_n)$ using the integers:

$$i_{j+1} = (ai_j + c) \pmod{m}, \quad j = 0, \ldots, n \tag{5.15}$$

where $m, a, c \in \mathbb{N}$ are positive integer numbers, called **modulus, multiplicator and increment**, while the starting term, i_0, is a non-negative integer called the **seed** of the generator; finally the u_j are obtained as $u_j = i_j/(m - 1)$. In the Table 5.1 we report typical values for the parameters m, a, c.

The reader could feel a bit confused now, since we have claimed independence while actually obtaining the sequence applying a deterministic (and very simple!) function to a given number to obtain the following one. This is the reason for the choices of the parameters in the given table, providing the conditions for the data $(u_1, \ldots, u_n)$ to be modeled by independent random variables. Statistical and numerical studies have shown that such choices of parameters make the model very accurate, in the sense discussed in the chapter about statistics.

The other important point is the seed i_0: it can be chosen to be equal to any non-negative integer number. If a program is used twice with the same seed, it gives exactly the same output. Actually the seed can be thought as the point $\omega \in \Omega$ in some abstract probability space $(\Omega, \mathcal{F}, P)$ determining the output of the "experiment":

$$(u_1, \ldots, u_n) = (U_1(\omega), \ldots, U_n(\omega)) \tag{5.16}$$

Table 5.1 parameters m, a, c of the LCG

Source	m	a	c
gclib	2^{31}	1103515245	12345
Numerical Recipes	2^{32}	1664525	1013904223
java.util.Random	2^{48}	25214903917	11

5.4 Simulation of Normal Random Variables

We have thus learned that, with a very simple algorithm, we can **sample** the uniform distribution in $(0, 1)$. What about other probability densities? In general, this problem is highly non trivial. Nevertheless, there are some situations allowing to solve the problem in a simple and elegant way, relying on transformations between random variables. Due to the outstanding importance of normal random variables, our first example is the sampling the density $N(0, 1)$:

$$p(x) = \frac{1}{\sqrt{2\pi}} \exp\left(-\frac{x^2}{2}\right) \tag{5.17}$$

We will exploit the transformation law for densities:

$$p_Y(\mathbf{y}) = p_X(g^{-1}(\mathbf{y})) \left| \det(J_{g^{-1}}(\mathbf{y})) \right| \tag{5.18}$$

valid whenever $Y = g(X)$, g being a diffeomorphism between open subsets of $\mathbb{R}^d$. $J_{g^{-1}}(\mathbf{y})$ is the Jacobian matrix of the inverse g^{-1}.

We specialize the transformation law to the special case in two dimensions $X = (U_1, U_2)$, U_1, U_2 being **independent uniform in** $(0, 1)$ and:

$$g(u_1, u_2) = \left(\sqrt{-2\log(u_1)} \cos(2\pi u_2), \sqrt{-2\log(u_1)} \sin(2\pi u_2) \right) \tag{5.19}$$

We let $Y = (Y_1, Y_2)$ and evaluate its density. The inverse if g is simply checked to be:

$$g^{-1}(y_1, y_2) = \left(\exp\left(-\frac{y_1^2 + y_2^2}{2}\right), \frac{1}{2\pi} \arctan\left(\frac{y_2}{y_1}\right) \right) \tag{5.20}$$

while its Jacobian is given by:

$$J_{g^{-1}}(y_1, y_2) = \begin{pmatrix} -y_1 \exp\left(-\frac{y_1^2 + y_2^2}{2}\right) & -y_2 \exp\left(-\frac{y_1^2 + y_2^2}{2}\right) \\ -\frac{y_2/y_1^2}{2\pi (1 + y_2^2/y_1^2)} & \frac{1/y_1}{2\pi (1 + y_2^2/y_1^2)} \end{pmatrix} \tag{5.21}$$

The determinant is:

$$\det\left(J_{g^{-1}}(y_1, y_2)\right) = -\frac{(1 + y_2^2/y_1^2)}{2\pi (1 + y_2^2/y_1^2)} \exp\left(-\frac{y_1^2 + y_2^2}{2}\right) \tag{5.22}$$

implying that $Y = (Y_1, Y_2) = g(U_1, U_2)$ has density:

$$p_Y(y_1, y_2) = \frac{1}{2\pi} \exp\left(-\frac{y_1^2 + y_2^2}{2}\right) \tag{5.23}$$

which means that Y_1 and Y_2 are **independent standard normal random variables**.

Thus, in practice, it is possible to use a random number generator twice, obtaining two numbers (u_1, u_2), and to apply the following **Box-Muller formula**:

$$y = \sqrt{-2\log(u_1)}\cos(2\pi u_2) \qquad (5.24)$$

to obtain a number sampling a standard normal random variable.

Remark 5.4 If the employed random number generator yields numbers in $[0, 1)$, the Box-Muller formula has to be modified by changing $u_1 \rightsquigarrow 1 - u_1$.

5.5   The Inverse Cumulative Distribution Function

We present now another very important example of the possibility of sampling one-dimensional random variables given a random number generator. Let's consider a given probability density $p(x)$ on $\mathbb{R}$ and let $F(x) = \int_{-\infty}^{x} dy\, p(y)$, the cumulative distribution function of a random variable having $p(x)$ as probability density. We work under the hypothesis that there exists an interval (α, β), $-\infty \leq \alpha < \beta \leq +\infty$ such that $p(x) > 0$ for $x \in (\alpha, \beta)$ and $p(x) = 0$ outside that interval. $F(x)$ is thus strictly increasing on (α, β) and its values lie in $[0,1]$. We define now $Y = F^{-1}(U)$ where U is uniform in $(0, 1)$. The key point is that the cumulative distribution of Y coincides with $F(x)$, in fact:

$$F_Y(y) = P(Y \leq y) = P(F^{-1}(U) \leq y) = P(U \leq F(y)) = F(y) \qquad (5.25)$$

and thus:

$$p_Y(y) = p(y) \qquad (5.26)$$

This means that we can sample any one-dimensional probability density $p(x)$ using a random number generator if we are able to evaluate F^{-1}: the generator provides a realization of a uniform random variable U, and, if we apply F^{-1}, we obtain a realization of a random variable with density $p(x)$.

$$\omega \rightsquigarrow U(\omega) \rightsquigarrow F^{-1}(U(\omega)) \qquad (5.27)$$

Example 5.1 If we wish to sample the **lorentzian** probability density:

$$p(x) = \frac{1}{\pi}\frac{\Gamma}{\Gamma^2 + x^2} \qquad (5.28)$$

we evaluate:

$$F(x) = \int_{-\infty}^{x} dy\, p(y) = \frac{1}{\pi}\arctan\left(\frac{x}{\Gamma}\right) + \frac{1}{2} \qquad (5.29)$$

We know that, if U is uniform in $(0, 1)$:

$$Y = F^{-1}(U) = \Gamma \tan\left(\pi\left(U - \frac{1}{2}\right)\right) \tag{5.30}$$

has density $p(x)$.

Example 5.2 If $p(x)$ is the **exponential density** with parameter λ:

$$p(x) = \lambda \exp(-\lambda x)\,1_{(0,+\infty)}(x), \quad \lambda > 0 \tag{5.31}$$

we evaluate:

$$F(x) = \int_{-\infty}^{x} dy\, p(y) = (1 - \exp(-\lambda x))\,1_{(0,+\infty)}(x) \tag{5.32}$$

so that, if U is uniform in $(0, 1)$:

$$Y = F^{-1}(U) = -\frac{1}{\lambda}\log(1 - U) \tag{5.33}$$

has density $p(x)$.

5.6 Discrete Random Variables

We discuss now the typical situation of the sampling a discrete probability density $p(x)$, non-zero only in the discrete set $\{x_1, \ldots, x_n\}$, which we assume to be finite. The typical tool we can use is the following: we define $q_0 = 0$, $q_1 = p(x_1)$, $q_2 = p(x_1) + p(x_2)$, $q_{m-1} = p(x_1) + p(x_2) + \cdots + p(x_{m-1})$ and, finally, $q_n = 1$. We have naturally $0 = q_0 < \cdots < q_n = 1$. If U is uniform in $(0, 1)$, we define:

$$Y = x_j, \quad if \quad q_{j-1} \leq U < q_j \tag{5.34}$$

Y is clearly a discrete random variable and has precisely the discrete density $p(x)$, as follows from the following simple calculation:

$$P(Y = x_j) = P(q_{j-1} \leq U < q_j) = q_j - q_{j-1} = p(x_j) \tag{5.35}$$

In practice, this result yields a very simple algorithm: we generate a random number u and find out:

$$j = \min\{n \mid \sum_{i=1}^{n} p(x_i) > u\} \tag{5.36}$$

Intuitively, this sampling scheme resembles an unfair roulette: the idea is to divide the interval $(0, 1)$ into n sub-intervals, the j-th having length equal to $p(x_j)$; then we sample $p(x)$ choosing x_j if a random number u falls inside the j-th sub-interval.

5.7 The Metropolis Algorithm

In the previous sections we have presented some tools to sample random variables, once a random number generator is available. Unfortunately, these tools are, in general, not useful in the multidimensional case, which is the most common field of application of Monte Carlo methods.

For example, in classical statistical mechanics, an extremely interesting topic is the possibility of sampling the Boltzmann weight of a classical fluid in thermal equilibrium at temperature $T = 1/K_B\beta$:

$$p(\mathbf{r}_1, \ldots, \mathbf{r}_N) = \frac{\exp\left(-\beta \sum_{i<j} v\left(|\mathbf{r}_i - \mathbf{r}_j|\right)\right)}{\mathcal{Z}} \qquad (5.37)$$

$v(r)$ being the interatomic potential. Another example is the celebrated Ising model, describing a collection of magnetic moments, *spins* $(\sigma_1, \ldots, \sigma_N)$, $\sigma_i = \pm 1$ on a lattice. The equilibrium properties, at temperature $T = 1/\beta$ and at the presence of a magnetic field B, are described by the probability density:

$$p(\sigma_1, \ldots, \sigma_N) = \frac{\exp\left(-\beta\left(-\sum_{\langle i,j \rangle} J\sigma_i\sigma_j - B\sum_i \sigma_i\right)\right)}{Z} \qquad (5.38)$$

$J > 0$ describing a ferromagnetic coupling between the nearest neighbours spins (the symbol $\langle i, j \rangle$ indicates that the summation is restricted to nearest neighbours).

Such probability densities can be sampled using the Metropolis algorithm, relying on the Metropolis theorem we have proved in the chapter about Markov chains. The basic idea is to build up a Markov chain, or more precisely a regular transition matrix, that has the desired probability density as its invariant law. It is then possible to choose any initial state and the Markov chain will converge to the desired law, in the sense discussed in the chapter about Markov chains.

We find useful to present the algorithm using the Ising model as a guiding example.

5.7.1 Monte Carlo Simulation of the Ising Model

The Ising model is certainly the most thoroughly studied model in statistical physics. It is a model of a magnet. The essential premise behind it is that the magnetism of a bulk material is made up of the combined magnetic dipole moments of many atomic spins within the material. The model postulates a lattice, typically an hyper cubic lattice $\mathbb{Z}^d$, with a magnetic dipole or spin on each site. In the Ising model these spins assume the simplest form possible, which consists of scalar variables σ_i which can take only two values ± 1, representing up-pointing or down-pointing dipoles of unit magnitude.

In a real, magnetic material the spins interact, and the Ising model mimics this by including terms in the Hamiltonian proportional to products $\sigma_i \sigma_j$ of the spins. In the simplest case, the interactions involves only nearest-neighbors spins and are all of the same strength, denoted by J (which has the dimensions of an energy), and the Hamiltonian is defined by:

$$\mathcal{H}(\sigma_1 \ldots \sigma_N) = -J \sum_{\langle ij \rangle} \sigma_i \sigma_j - B \sum_i \sigma_i \qquad (5.39)$$

where the notation $\langle ij \rangle$ indicates that the sum runs over nearest neighbours. The minus signs here are conventional. They merely dictate the choice of sign for the interaction parameter J and the external magnetic field B. A positive value for J signals a ferromagnetic coupling.

We want to study the equilibrium distribution of the model for N spins at temperature $T = 1/\beta$, that is the probability density:

$$\pi(\sigma_1 \ldots \sigma_N) = \frac{1}{Z(\beta, B, N)} \exp\left(-\beta \mathcal{H}(\sigma_1 \ldots \sigma_N)\right) \qquad (5.40)$$

defined on the state space $E = \{(\sigma_1 \ldots \sigma_N), \ \sigma_i = \pm 1\}$ containing 2^N possible configurations of spins. The denominator $Z(\beta, B, N)$ is the partition function of the model:

$$Z(\beta, B, N) = \sum_{(\sigma_1 \ldots \sigma_N) \in E} \exp\left(-\beta \mathcal{H}(\sigma_1 \ldots \sigma_N)\right) \qquad (5.41)$$

We focus our attention on the simplest situation: the one-dimensional case. We adopt also periodic boundary conditions, defining:

$$\mathcal{H}(\sigma_1 \ldots \sigma_N) = -J \sum_{i=1}^{N} \sigma_i \sigma_{i+1} - B \sum_{i=1}^{N} \sigma_i, \quad \sigma_{N+1} \equiv \sigma_1 \qquad (5.42)$$

We will learn now how to implement the Metropolis algorithm to sample π. We will use the notation $\sigma \equiv (\sigma_1 \ldots \sigma_N)$ for the states, i.e. the configurations of the spins.

5.7.1.1 The Algorithm

The algorithm consists in the iteration of some steps, translating into a practical algorithm the choice of an initial law and the sampling the law of a Markov chain

$\{\Sigma_n\}_n$, with a transition matrix of the form:

$$\mathcal{P}_{\sigma\to\sigma'} = \mathcal{T}_{\sigma\to\sigma'}\,\mathcal{A}_{\sigma\to\sigma'} \tag{5.43}$$

where $\mathcal{T}$ is any symmetric and irreducible transition matrix, while, as we learned in the previous chapter, the *acceptance probability*, is:

$$\mathcal{A}_{\sigma\to\sigma'} = \min\left(1, \frac{\pi(\sigma')}{\pi(\sigma)}\right) \tag{5.44}$$

Metropolis theorem ensures that:

$$\lim_{n\to+\infty} P\left(\Sigma_n = \sigma\right) = \pi(\sigma) \tag{5.45}$$

The steps are the following:

1. **Inizialization**. We start choosing an arbitrary initial state, that is an initial configuration for the spins σ_0. For example, we can make all spins up-pointing.
2. **Trial move**. We propose a move $\sigma_0 \rightsquigarrow \sigma_{trial}$, randomly choosing a spin and flipping it.
3. **Acceptation**. If we have flipped, say, the jth spin, we have $\sigma_0 = (\sigma_1, \ldots, \sigma_j, \ldots, \sigma_N)$ and $\sigma_{trial} = (\sigma_1, \ldots, -\sigma_j, \ldots, \sigma_N)$.
   We evaluate the number:

$$w = \frac{\pi(\sigma_{trial})}{\pi(\sigma_0)} = \exp\left(-\beta\left[\mathcal{H}(\sigma_{trial}) - \mathcal{H}(\sigma_0)\right]\right) \tag{5.46}$$

   that is:

$$w = \exp\left(-2\beta B\sigma_j - 2\beta J\sigma_j(\sigma_{j-1} + \sigma_{j+1})\right) \tag{5.47}$$

   Then, we generate a uniform random number $r \in (0, 1)$ and:

   a. if $r \leq w$, we accept the move, defining $\sigma_1 = \sigma_{trial}$;

   b. if $r > w$, we reject the move, defining $\sigma_1 = \sigma_0$.

4. **Iteration**. Then, we use σ_1 as the new starting point, and go back to point 2.

We proceed with the process for, at least, $M \simeq 10^5 - 10^6$ Monte Carlo steps.

Remark 5.5 We stress a very important point: the implementation of the Metropolis algorithm does not require the evaluation of the partition function

$$Z(\beta, B, N) = \sum_{(\sigma_1 \ldots \sigma_N) \in E} \exp\left(-\beta \mathcal{H}(\sigma_1 \ldots \sigma_N)\right) \tag{5.48}$$

since it involves only the ratio of values of the probability density. In the special case of the one-dimensional Ising model the partition function can be evaluated analytically, but this is an exception: there is only an handful of problems in statistical physics that allow for an analytical solution.

The Metropolis algorithm builds up a realization of a Markov chain with state space made of all the possible configurations of the spins:

$$\sigma_0 \rightsquigarrow \sigma_1 \rightsquigarrow \cdots \rightsquigarrow \sigma_n \rightsquigarrow \ldots \tag{5.49}$$

From the theory of Markov chains we know that, for n large enough, the $\{\sigma_n\}$ sample the probability π.

As the spins "move" exploring the state space, we can perform measurements on the system, actually like an experimentalist. For example, suppose we wish to evaluate the average magnetization:

$$\mathcal{M}(B, \beta) = \sum_{\sigma} m(\sigma) \pi(\sigma) \tag{5.50}$$

where:

$$m(\sigma) = \frac{1}{N} \sum_{i=1}^{N} \sigma_i \tag{5.51}$$

The basic idea is to estimate it computing an empirical mean over the random walk:

$$\mathcal{M}(B, \beta) \simeq \frac{1}{M} \sum_{n=0}^{M} m_n, \quad m_n = m(\sigma_n) \tag{5.52}$$

Incidentally, we stress that it is neither mandatory nor in general advisable to use all the steps of the Markov chain to perform the measurements; on the contrary, in the realm of Monte Carlo simulations, the evaluation of a quantity on a configuration happens to be much more computationally expensive than producing a new configuration (actually this is not the case of the magnetization in Ising model), making more convenient to wait some time before measuring again the same quantity. This procedure is called **sparse averaging**; we will come back to this point in a while. By now it is enough to say that the summation (5.52) involves a set of samples of the magnetization, m_n, measured at evenly-spaced "times" n, while the spins are moving.

Although the estimation (5.52) can be justified relying on ergodic theorems, in practice one wishes to employ the central limit theorem, in order to provide an estimation of the magnetization of the system together with a statistical uncertainty.

In principle this is not allowed, since the σ_n sample the desired probability density only asymptotically, for large n and, moreover, they are neither independent nor identically distributed, being realizations of the steps of a Markov chain.

Some empirical strategies are commonly adopted to recover the conditions to apply, at least approximately, the central limit theorem.

The first is **equilibration**: the first steps of the random walk, when the distribution of the sampled Markov chain has not yet reached its limit law π, are discarded. This is done empirically, monitoring, for example, the instantaneous value of the magnetization itself as a function of the number of steps. It can be useful to perform more than one simulation, changing the initial configuration: after a transient, the values of the magnetization begin to fluctuate around a value independent of the initial condition. This transient corresponds to the steps we have to neglect for our calculations. It is interesting to observe that such equilibration transient resembles the typical thermalization of physical systems approaching a steady equilibrium state.

Let's consider now the correlations among the measurements. The summation (5.52) is a realization of the random variable:

$$S_M = \frac{1}{M} \sum_{n=1}^{M} M_n \tag{5.53}$$

where, after the equilibration transient, the M_n have density π but are not independent. Naturally $E[S_M] = \mathcal{M}(B, \beta)$. Let's look at the variance:

$$Var(S_M) = E\left[(S_M - \mathcal{M}(B,\beta))^2\right] =$$
$$= \frac{1}{M^2} E\left[\sum_{n,l=1}^{M} (M_n - \mathcal{M}(B,\beta))(M_l - \mathcal{M}(B,\beta))\right] = \tag{5.54}$$
$$= \frac{1}{M^2} \sum_{n,l=1}^{M} Cov(M_n, M_l)$$

As a consequence of the homogeneity of the Markov chain, the covariance $Cov(M_n, M_l)$ depends only on the time $t = n - l$. We define the autocorrelation function of the magnetization:

$$C_\mathcal{M}(t) = Cov(M_n, M_{n+t}) \tag{5.55}$$

We thus write:

$$Var(S_M) = \frac{1}{M^2} \sum_{n,l=1}^{M} Cov(M_n, M_l) =$$

$$\simeq \frac{C_M(0)}{M} + \frac{1}{M^2} \sum_{n=1}^{M} \sum_{t \neq 0} C_M(t) \tag{5.56}$$

where the last member is approximate since the summation over t is extended to the whole $\mathbb{Z} - \{0\}$. To be consistent with such approximation, which somehow forgets the initial time and the finite number of measurements, we assume that $C_M(t) = C_M(|t|)$. We find the approximate result:

$$Var(S_M) \simeq \frac{C_M(0)}{M} + \frac{1}{M^2} \sum_{n=1}^{M} \sum_{t \neq 0} C_M(t) =$$

$$= \frac{C_M(0)}{M} \left(1 + 2 \sum_{t \neq 0} \frac{C_M(t)}{C_M(0)} \right) \tag{5.57}$$

The number:

$$\left(1 + 2 \sum_{t \neq 0} \frac{C_M(t)}{C_M(0)} \right) \stackrel{def}{=} 2\tau_M \tag{5.58}$$

defines the **autocorrelation time** of the magnetization. The name derives from the expected large time behavior:

$$\frac{C_M(t)}{C_M(0)} \simeq \exp\left(-\frac{t}{\tau_M} \right) \tag{5.59}$$

We observe that:

$$Var(S_M) \simeq \frac{C_M(0)}{M_{eff}} \tag{5.60}$$

where the effective number of data is:

$$M_{eff} = \frac{M}{2\tau_M} \tag{5.61}$$

Comparing this result with the one about independent and identically distributed random variables we have presented when dealing with the central limit theorem, the following procedure is quite natural: we use all the measured data to estimate the autocorrelation function $C_M(t)$, using the formula:

$$\frac{1}{M-t}\sum_{i=1}^{M-t} m_i m_{i+t} - \frac{1}{M-t}\sum_{i=1}^{M-t} m_i \frac{1}{M-t}\sum_{i=0}^{M-t} m_{i+t} \qquad (5.62)$$

and subsequently estimating the autocorrelation time. Finally, it is advisable to keep, in the summation (5.52), only measurements obtained every $2\tau_\mathcal{M}$ measurements (**sparse averaging**). In this way, we can safely deal with the measurements as if they were independent.

As we have learned in the chapter about statistics, it is desirable to have normal samples, in order to estimate confidence intervals on a rigorous basis. This is achieved performing **data blocking**, which actually corresponds to perform many different independent simulations. The simulation is divided in large (several autocorrelation times) blocks; for each block an estimation of $\mathcal{M}(B, \beta)$ is provided. If the blocks are large enough, we may rely on the central limit theorem and interpret each estimation as a realization of a normal random variable; we can thus use mathematical statistics to provide a confidence interval for the average magnetization of the system.

5.7.1.2 Analytic Results

A reader wishing to implement a Monte Carlo simulation can find useful to compare his/her results with the analytic results for the Ising model in one-dimension. The explicit solution relies on the observation that the partition function can be written as:

$$Z(\beta, B, N) = Tr\left(\mathcal{T}^N\right) \qquad (5.63)$$

where:

$$\mathcal{T} = \begin{pmatrix} e^{\beta(J+B)} & e^{-\beta J} \\ e^{-\beta J} & e^{\beta(J-B)} \end{pmatrix} \qquad (5.64)$$

Thus, we have:

$$Z(\beta, B, N) = \tau_+^N + \tau_-^N \qquad (5.65)$$

$\tau_\pm$ being the eigenvalues of the matrix $\mathcal{T}$. explicitly:

$$\tau_\pm = e^{\beta J}\left(\mathrm{Ch}(\beta B) \pm \sqrt{\mathrm{Sh}(\beta B)^2 + e^{-4\beta J}}\right) \qquad (5.66)$$

It is to possible use this expressions to evaluate analytically the magnetization:

$$M(\beta, B) = \langle\sigma_i\rangle = \frac{1}{N\beta}\frac{\partial \log(Z(\beta, B, N))}{\partial B} \qquad (5.67)$$

The expression for finite N is quite cumbersome, but becomes simple in the thermodynamic limit $N \to +\infty$:

$$M(\beta, B) = \langle \sigma_i \rangle = \frac{\text{Sh}(\beta B) + \frac{\text{Sh}(\beta B)\text{Ch}(\beta B)}{\sqrt{\text{Sh}^2(\beta B) + e^{-4\beta J}}}}{\text{Ch}(\beta B) + \sqrt{\text{Sh}^2(\beta B) + e^{-4\beta J}}} \qquad (5.68)$$

5.7.2 Monte Carlo Simulation of a Classical Simple Liquid

Another very important example of application of Monte Carlo simulations is the study of a classical fluid in thermal equilibrium at temperature $T = 1/K_B\beta$. A typical model of the two-body interaction among the particle in the fluid, very accurate in the case of noble gases, is the following Lennard-Jones potential, depending only on the inter particle distance:

$$v(r) = 4\varepsilon \left\{ \left(\frac{\sigma}{r}\right)^{12} - \left(\frac{\sigma}{r}\right)^6 \right\} \qquad (5.69)$$

The r^{-12} term describes hard-core interaction, while the attractive r^{-6} term represents Van der Waals weak induced dipole attraction. The phenomenological parameters ε and σ depend on the particular noble gas one wishes to study.

We are interested in the bulk physics of the system: we fix the particles density ρ and consider a cubic region, say $B \subset \mathbb{R}^3$, of volume $V = L^3$ with $\rho = \frac{N}{V}$. Furthermore, we imagine that the whole $\mathbb{R}^3$ is covered by identical replicas of the simulation box B.

In order to be consistent with such a picture, we let $\mathcal{R} = (\mathbf{r}_1, \ldots, \mathbf{r}_N)$ denote the configuration of the fluid inside B and we write the potential energy of the N particles moving inside B and interacting with all the *images* of the particles in the replicas of B, in the form:

$$\mathcal{V}(\mathcal{R}) = \sum_{1 \leq i < j \leq N} \tilde{v}\left(\mathbf{r}_{i,j}\right), \quad \mathbf{r}_{i,j} = \mathbf{r}_i - \mathbf{r}_j \qquad (5.70)$$

with:

$$\tilde{v}(\mathbf{r}) = \begin{cases} v\left(|\mathbf{r} - L \, \text{nint}\left(\frac{\mathbf{r}}{L}\right)|\right) - v(r_c), & |\mathbf{r} - L \, \text{nint}\left(\frac{\mathbf{r}}{L}\right)| \leq r_c \\ 0, & |\mathbf{r} - L \, \text{nint}\left(\frac{\mathbf{r}}{L}\right)| > r_c \end{cases} \qquad (5.71)$$

where $r_c \leq \frac{L}{2}$ is a cutoff radius, larger than the interaction range. The redefinition of the distances $r_{ij} = |\mathbf{r}_i - \mathbf{r}_j| \rightarrow |\mathbf{r}_{ij} - L \, \text{nint}\left(\frac{\mathbf{r}_{ij}}{L}\right)|$, where nint denotes the nearest integer, is meant to be consistent with the picture of infinite replicas of the simulation box, where the images of the particles reproduce the motion of the particles themselves. The number $|\mathbf{r}_{ij} - L \, \text{nint}\left(\frac{\mathbf{r}_{ij}}{L}\right)|$ is the distance between particle i and the particle j in the box whenever the cartesian components of $\mathbf{r}_{ij}$ are smaller than $L/2$, otherwise it is the distance between particle i and the image of particle j in neighboring boxes closest to the particle i. We stress that it is important that the potential (5.71) is short range, making only the images in the nearest replicas interact with the particles in the box B (**minimum image convention**). This is the reason why the potential is cut at r_c. Moreover, the potential is shifted by a quantity $v(r_c)$, in order to avoid a discontinuity at the point r_c. Although this shift does not have any effect in Monte Carlo simulations, it is a common practice to introduce it. The potential (5.71) is usually called the **cut and shifted** Lennard-Jones potential. We will not present here the implications of such choice in the definition of the interatomic potential: an interested reader see, for example, the excellent textbook [3, 4].

Now, once fixed the mechanical model of the classical fluid, in thermal equilibrium the configurational probability density has the form:

$$p(\mathcal{R}) = \frac{\exp\left(-\beta V(\mathcal{R})\right)}{\int_{B^N} d\mathcal{R}' \exp\left(-\beta V(\mathcal{R}')\right)} \tag{5.72}$$

Once again, it is possible to use the Metropolis algorithm to sample the Gibbs weight (5.72). To do this, we have to build up a Markov chain whose state space is B^N:

$$\{\mathbf{R_n}\}_{n=0,1,2,\ldots} \tag{5.73}$$

starting from a given initial configuration $\mathbf{R_{n=0}} = \mathcal{R}_0$, typically chosen sitting the particles on the sites of a crystalline lattice. According to Metropolis theorem, we choose the transition matrix of the form:

$$\mathcal{P}_{\mathcal{S} \to \mathcal{R}} = \mathcal{T}_{\mathcal{S} \to \mathcal{R}} \, \mathcal{A}_{\mathcal{S} \to \mathcal{R}}, \quad \mathcal{A}_{\mathcal{S} \to \mathcal{R}} = \min\left(1, \frac{p(\mathcal{R})}{p(\mathcal{S})}\right) \tag{5.74}$$

where $\mathcal{T}_{\mathcal{S} \to \mathcal{R}}$ is a symmetric and irreducible transition matrix. In general the particles are moved one at the time; this means that $\mathcal{T}_{\mathcal{S} \to \mathcal{R}}$ is chosen to be non zero only if $\mathcal{R}$ and $\mathcal{S}$ differ only in the position vector of one particle. In such case, we choose to move a particle randomly uniformly inside a cube centered on the actual position, that is:

$$\mathcal{T}_{(\mathbf{s}_1,\ldots,\mathbf{s}_i,\ldots\mathbf{s}_N) \to (\mathbf{s}_1,\ldots,\mathbf{r}_i,\ldots\mathbf{s}_N)} = \frac{1}{N} \frac{1}{\Delta^3} \tag{5.75}$$

if $\mathbf{r}_i$ lies inside a cube of edge 2Δ centered in $\mathbf{s}_i$; otherwise, the transition matrix will be identically zero. The free parameter Δ is typically chosen *a posteriori*, monitoring the acceptation rate of the Metropolis moves, and fixing it in such a way that nearly the 50% of the moves are accepted.

In practice, it is very simple to build up the simulation: starting from the initial configuration, we propose a random displacement of a particle, and accept/reject the move according to the ration $\frac{p(\mathcal{R})}{p(\mathcal{S})}$, exactly in the same way as in the case of the Ising model. Iterating this for a suitable number of Monte Carlo steps we can perform measurements on the fluid, evaluating average properties.

The simplest quantities that can be computed are the average potential energy:

$$\left\langle \frac{V}{N} \right\rangle = \frac{1}{N} \int_{B^N} d\mathcal{R} \, V(\mathcal{R}) \, p(\mathcal{R}) \tag{5.76}$$

and the virial:

$$\langle W \rangle = \int_{B^N} d\mathcal{R} \left(\sum_{i=1}^{N} \mathbf{r}_i \cdot \nabla_i V(\mathcal{R}) \right) p(\mathcal{R}) \tag{5.77}$$

giving access to the pressure of the fluid through the virial relation, which can be found in any textbook about Statistical Mechanics:

$$P = \rho k_B T - \frac{1}{3V} \langle W \rangle \tag{5.78}$$

Other very important properties that can be evaluated in a Monte Carlo simulation are the correlation functions among the positions of the particles. In standard textbooks on statistical mechanics, the reduced two-body density matrix is introduced:

$$\rho^{(2)}(\mathbf{r}, \mathbf{r}') = N(N-1) \int d\mathbf{r}_3 \dots d\mathbf{r}_N \, p\left(\mathbf{r}, \mathbf{r}', \mathbf{r}_3, \dots, \mathbf{r}_N\right) \tag{5.79}$$

The interpretation of this quantity is given by the joint probability density for two-particles, that is:

$$\int_{A \times A'} d\mathbf{r} d\mathbf{r}' \rho^{(2)}(\mathbf{r}, \mathbf{r}') \tag{5.80}$$

is the probability to find a particle in the region A and another particle in the region A'. Let's now evaluate the probability to find two particles with a relative distance between r and $r + \Delta r$:

$$\int_{r \leq |\mathbf{r} - \mathbf{r}'| \leq r + \Delta r} d\mathbf{r} d\mathbf{r}' \rho^{(2)}(\mathbf{r}, \mathbf{r}') = \int_{r \leq |\mathbf{r}| \leq r + \Delta r} d\mathbf{r} d\mathbf{r}' \rho^{(2)}(\mathbf{r} + \mathbf{r}', \mathbf{r}') \tag{5.81}$$

where we have simply changed variable $\mathbf{r} \to \mathbf{r} - \mathbf{r}'$. In the liquid phase, we expect that the reduced two-body density matrix $\rho^{(2)}(\mathbf{r}, \mathbf{r}')$ depends only on $|\mathbf{r} - \mathbf{r}'|$, so that we can write:

$$\int_{r \leq |\mathbf{r} - \mathbf{r}'| \leq r + \Delta r} d\mathbf{r} d\mathbf{r}' \rho^{(2)}(\mathbf{r}, \mathbf{r}') = V \int_{r \leq |\mathbf{r}| \leq r + \Delta r} d\mathbf{r} \rho^{(2)}(\mathbf{r}) =$$
$$= V \int_0^\pi d\theta \sin\theta \int_0^{2\pi} d\varphi \int_r^{r+\Delta r} ds s^2 \rho^2 \, g(s) \tag{5.82}$$

where we have introduced the **radial distribution function**:

$$g(r) = \frac{\rho^{(2)}(r)}{\rho^2} \tag{5.83}$$

In we choose Δr small enough so that we can neglect the variations of $g(r)$ in the interval $[r, r + \Delta r]$, we may write:

$$\int_{r \leq |\mathbf{r} - \mathbf{r}'| \leq r + \Delta r} d\mathbf{r} d\mathbf{r}' \rho^{(2)}(\mathbf{r}, \mathbf{r}') \simeq g(r) \frac{4\pi \rho N}{3} \left((r + \Delta r)^3 - r^3 \right) \tag{5.84}$$

We can rewrite the above relation in the following way, keeping into account also our model of the system in periodic boundary conditions:

$$g(r) = \mathcal{N}(r, \Delta r) \int d\mathcal{R} \left(\sum_{i \neq j=1}^{N} 1_{[r, r + \Delta r]} \left(|\mathbf{r}_{ij} - L \, \text{nint} \left(\frac{\mathbf{r}_{ij}}{L} \right)| \right) \right) p(\mathcal{R}) \tag{5.85}$$

where the normalization is:

$$\mathcal{N}(r, \Delta r) = \frac{1}{\frac{4\pi \rho N}{3} \left((r + \Delta r)^3 - r^3 \right)} \tag{5.86}$$

From the expression (5.85) it is evident that we can evaluate $g(r)$ during the simulation: we have simply to build up an histogram of the distances among particles.

5.8 Further Readings

The field of Monte Carlo integration and numerical simulations is very vast. Our focus is on applications within statistical mechanics. Readers interested in the foundations of statistical mechanics can see, for example, [1, 2]. For readers more interested in the applications of numerical simulations, we recommend the classical textbooks [3, 4]. Finally, more information about random number generation and Monte Carlo simulation can be found, for example, in [5].

Problems

5.1 Buffon needle
Perhaps the earliest documented use of random sampling to find the solution to an integral is that of Comte de Buffon. In 1777 he described the following experiment: a needle of length L is thrown at random onto a horizontal plane ruled with straight lines a distance d apart ($d > L$). What is the probability P that the needle will intersect one of these lines? Describe an algorithm to evaluate π starting from this experiment.

5.2 Uniform law inside the unit circle

Consider a random variable X uniform inside the unit circle, with density:

$$p_X(\mathbf{x}) = \begin{cases} \frac{1}{\pi}, & |\mathbf{x}| \leq 1 \\ 0, & otherwise \end{cases} \tag{5.87}$$

Show that X can be sampled in the following way: using a random number generator, we continue generating points within the square $\{\mathbf{x} = (x, y) \in \mathbb{R}^2 \mid -1 \leq x \leq 1, \ -1 \leq y \leq 1\}$, and keep only the ones falling inside the circle.

5.3 Another way to sample the normal distribution

Consider the random variable X of the previous exercise, and define the two dimensional random variable $Y = g(X)$, where:

$$g(\mathbf{x}) = \mathbf{x} \sqrt{\frac{-2 \log\left(|\mathbf{x}|^2\right)}{|\mathbf{x}|^2}} \tag{5.88}$$

show that $Y = (Y_1, Y_2)$ where Y_1 and Y_2 are independent standard normal random variables.

5.4 Simulation of a poisson process

Suppose we wish simulate a radioactive decay process. Imagine we switch up a timer, and we let T_1 be the instant of the first decay, T_2 of the second, and so on. The number of decays in the interval $[0, t]$ is given by:

$$N_t = \sup\{n \mid T_n \leq t\} \tag{5.89}$$

In this problem we will learn the law of N_t under quite natural assumptions and how to sample it.

We will write:

$$T_n = \Delta T_1 + \Delta T_2 + \cdots + \Delta T_n \tag{5.90}$$

where $\Delta T_i = T_i - T_{i-1}$ is the inter-arrival time. We assume that the random variables ΔT_i are independent, identically distributed, and follow an exponential law with parameter λ, that is:

$$p_{\Delta T}(x) = \begin{cases} \lambda \exp(-\lambda x), & x > 0 \\ 0, & x \leq 0 \end{cases} \tag{5.91}$$

Show that N_t follows a Poisson law with parameter λt, that is:

$$P(N_t = k) = \frac{(\lambda t)^k}{k!} \exp(-\lambda t) \tag{5.92}$$

Using this result, describe an algorithm to sample N_t for any t, that is to simulate the Poisson process.

5.5 Methods to reduce the variance

Suppose we wish to evaluate the integral:

$$I = \int_0^1 dx \, \exp(x) \quad (= e - 1) \tag{5.93}$$

using a Monte Carlo method.

First, we use a random number generator to generate $x_1, \ldots, x_n$ sampling a uniform random variable U and estimate I as $I = E[f(U)]$ where $f(u) = \exp(u)$. Compute $Var(f(U))$, controlling the precision of the estimation.

Then we use an **importance sampling** technique, rewriting:

$$I = \int_0^1 dx \, \exp(x) = \int_0^1 dx \, \left(\frac{3}{2} \frac{\exp(x)}{1+x} \right) \frac{2}{3}(1 + x) \tag{5.94}$$

and generating $x_1, \ldots, x_n$ sampling a random variable X with density $p_X(x) = \frac{2}{3}(1 + x) \, 1_{(0,1)}(x)$ to estimate I as $I = E[g(X)]$ where $g(x) = \frac{3}{2} \frac{\exp(x)}{1+x}$. Compute $Var(g(X))$.

Now, we adopt another strategy, the **Antithetic Variates** technique, writing:

$$I = \int_0^1 dx \, \exp(x) = \int_0^1 dx \, \frac{1}{2} (\exp(x) + \exp(1 - x)) \tag{5.95}$$

and we use again a random number generator to generate $x_1, \ldots, x_n$ sampling a uniform random variable U, but now we estimate I as $I = E[h(U)]$ where $h(u) = \frac{1}{2}(\exp(u) + \exp(1 - u))$. Compute $Var(h(U))$.

Finally, we combine the two strategies writing:

$$I = \int_0^1 dx \, \exp(x) = \int_0^1 dx \, \frac{1}{2} \left(\frac{\exp(x) + \exp(1 - x)}{\pi(x)} \right) \pi(x) \tag{5.96}$$

where $\pi(x) = \frac{24}{25} \left(1 + \frac{1}{2} \left(x - \frac{1}{2} \right)^2 \right) 1_{(0,1)}(x)$. We generate $x_1, \ldots, x_n$ sampling a random variable X with density $\pi(x)$ and we estimate I as $I = E[h(X)]$ where $h(X) = \frac{1}{2} \left(\frac{\exp(x) + \exp(1 - x)}{\pi(x)} \right)$. Compute $Var(h(X))$.

5.6 A delicate situation

Suppose we wish to evaluate the integral:

$$I = \int_0^1 dx \, \frac{1}{\sqrt{x}} \quad (= 2) \tag{5.97}$$

using a Monte Carlo method.

Show that, if we use a random number generator to generate $x_1, \ldots, x_n$ sampling a uniform random variable U and estimate I as $I = E[f(U)]$ where $f(u) = \frac{1}{\sqrt{u}}$, we have $Var(f(U)) = +\infty$.

On the other hand, show that, if we use importance sampling, writing:

$$I = \int_0^1 dx \, \frac{1}{\sqrt{x}} = \int_0^1 dx \, \frac{x^{-1/2}}{\pi(x)} \pi(x) \tag{5.98}$$

with, for example, $\pi(x) = (1 - r) x^{-r} 1_{(0,1)}(x)$, for $0 < r < 1$, makes the variance finite allowing to perform the calculation sampling the density $\pi(x)$.

References

1. Huang, K.: Introduction to Statistical Physics. Taylor & Francis (2001)
2. Schwabl, F., Brewer, W.D.: Statistical Mechanics. Springer (2006)
3. Frenkel, D., Smit, B.: Understanding Molecular Simulations. Academic Press (1996)
4. Allen, M.P., Tildesley, D.J.: Computer Simulation of Liquids, 2nd ed. Oxford University Press (2017)
5. Gentle, J.E.: Random Number Generation and Monte Carlo Method. Paperback (2003)

Chapter 6
Brownian Motion

Abstract In this chapter we will introduce the celebrated brownian motion. Starting from the study of the randomly-driven motion of a pollen grain, first observed by the botanist Robert Brown, the brownian motion has become the cornerstone of the theory of stochastic processes in the continuum. The brownian motion is a sort of meeting point of several aspects of abstract mathematics, theoretical physics and real-world applications. We present first an heuristic description of it due to Einstein's genius, and then we provide a rigorous definition, taking the opportunity to introduce important definitions of stochastic processes theory. Afterwards, we define the transition probability, the generalization of the transition matrix of Markov chains. We also present the intriguing connection with Feynman's path integral and Quantum Mechanics: the brownian motion is the *free particle* motion in (imaginary time) quantum mechanics. Finally, we start exploring a deep connection existing between the theory of stochastic processes and the theory of partial differential equations: the brownian motion is related to the celebrated heat equation.

Keywords Brownian motion · Stochastic processes · Martingale property
Markov property · Heat equation

6.1 Introduction

Markov chains, presented in Chap. 4, belong to a wide class of mathematical objects, the stochastic processes, which we will introduce in the present chapter. Stochastic processes form the basis for the mathematical modeling of non deterministic time-dependent phenomena. We will begin focussing our attention on the randomly-driven motion of a pollen grain, the well-known brownian motion. Such motion, historically, has been the first stochastic process to be stated in rigorous mathematical terms. The brownian motion will represent our starting point for introducing the fundamental notions of stochastic processes. It will be introduced retracing the pioneering work of A. Einstein.

© Springer International Publishing AG, part of Springer Nature 2018   131
E. Vitali et al., *Theory and Simulation of Random Phenomena*, UNITEXT
for Physics, https://doi.org/10.1007/978-3-319-90515-0_6

6.2 Brownian Motion: A Heuristic Introduction

For the sake of simplicity, the one-dimensional case will be considered first, the generalization to higher dimensionality being straightforward. We will assume that in a given time interval τ a pollen grain, due to random collisions with water molecules, undergoes a random variation ζ in its position. We will denote through

$$\Phi(\tau, \zeta) \tag{6.1}$$

the probability density for the "jump" of the pollen grain, in such a way that

$$\int_a^b \Phi(\tau, \zeta)\, d\zeta \tag{6.2}$$

can be interpreted as the probability that the particle, being in any given position x at time t, is found in the interval $[x + a, x + b]$ at time $s = t + \tau$; we are assuming in particular that the *transition probability density* $\Phi(\tau, \zeta)$ depends only on the quantities $\tau = t - s$ and $\zeta = y - x$ (Fig. 6.1).

The function $\Phi(\tau, \zeta)$ remarkably resembles the transition matrix of a Markov chain, except for the fact that, in the present context, time and state space are continuous. We will assume that the following property:

$$\Phi(\tau, \zeta) = \Phi(\tau, -\zeta) \tag{6.3}$$

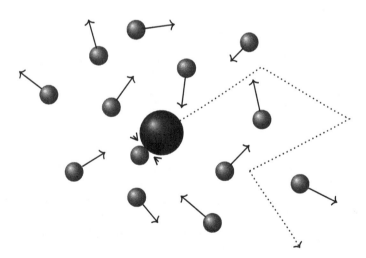

Fig. 6.1 The pollen grain (large sphere) moves inside the water, making random collisions with water molecules (small spheres)

holds, which expresses the fact that moving rightwards of leftwards is equally prob-
able, and implies that the "mean jump" vanishes:

$$\int_{\mathbb{R}} d\zeta \, \zeta \, \Phi(\tau, \zeta) = 0 \qquad (6.4)$$

Let now $p(x, t)$ denote the probability density for the position of the pollen grain at
time t. We will assume that the following very natural *continuity equation* holds:

$$p(x, t + \tau) = \int_{\mathbb{R}} d\zeta \, \Phi(\tau, \zeta) \, p(x - \zeta, t) \qquad (6.5)$$

expressing the fact that the position at time $s = t + \tau$ is the sum of the position at
time t and of the transition occurred between times t and s. If τ is *small*, provided that
$\Phi(\tau, \zeta)$ tends to 0 rapidly enough as $|\zeta| \to +\infty$, we can make Taylor expansions:

$$p(x, t + \tau) = p(x, t) + \tau \frac{\partial p}{\partial t}(x, t) + \dots \quad , \qquad (6.6)$$

$$p(x - \zeta, t) = p(x, t) - \zeta \frac{\partial \Phi}{\partial x}(x, t) + \frac{1}{2} \zeta^2 \frac{\partial^2 \Phi}{\partial^2 x}(x, t) + \dots \qquad (6.7)$$

Substituting the above expressions into the continuity equation (6.5) and retaining
non vanishing low-order terms only we get:

$$p(x, t) + \tau \frac{\partial p}{\partial t}(x, t) + \dots = p(x, t) + \frac{1}{2} \frac{\partial^2 p}{\partial^2 x}(x, t) \int_{\mathbb{R}} d\zeta \, \zeta^2 \Phi(\tau, \zeta) + \dots \qquad (6.8)$$

where we have taken into account the fact (6.4) that the mean transition is vanishing,
enabling us to retain terms of at most first order in τ and second in ζ, and that the
transition probability density is normalized to 1. Inspired by the microscopic theory
of diffusion, we are induced to interpret the quantity:

$$D = \frac{1}{2\tau} \int_{\mathbb{R}} d\zeta \, \zeta^2 \Phi(\tau, \zeta) = \frac{\sigma^2(\tau)}{2\tau} \qquad (6.9)$$

as the macroscopic diffusion coefficient; for this quantity to be *constant*, it is necessary to assume that $\sigma^2(\tau)$ is proportional to τ. Under this additionally hypothesis,
one arrives to the **diffusion equation**:

$$\frac{\partial p}{\partial t}(x, t) = D \frac{\partial^2 p}{\partial^2 x}(x, t) \qquad (6.10)$$

To complete the description, it is necessary to equip the diffusion equation (6.10) with an initial condition. A possible choice is:

$$p(x, 0) = \delta(x) \tag{6.11}$$

representing a particle at the origin of a suitable reference frame. Were that the case, the solution of (6.10) would be a gaussian probability density:

$$p(x, t) = \frac{1}{\sqrt{4\pi Dt}} \exp\left(-\frac{x^2}{4Dt}\right) \tag{6.12}$$

which coincides, by virtue of the continuity equation (6.5), with the transition probability density:

$$\Phi(\tau, \zeta) = \frac{1}{\sqrt{4\pi D\tau}} \exp\left(-\frac{\zeta^2}{4Dt}\right) \tag{6.13}$$

This expression will represent a natural starting point for the rigorous mathematical treatment of the brownian motion.

6.3 Stochastic Processes: Basic Definitions

The heuristic discussion of the preceding paragraph has put in evidence several remarkable aspects: first, the motion of the pollen grain is treated with probabilistic arguments. This implies that a probability space $(\Omega, \mathcal{F}, P)$ has to be introduced, on which it must be possible to define the random variables:

$$X_t : (\Omega, \mathcal{F}, P) \rightarrow \left(\mathbb{R}^d, \mathcal{B}\left(\mathbb{R}^d\right)\right) \tag{6.14}$$

with the interpretation of position of the pollen grain at time t. The *time parameter t* takes values on a suitable *time interval* $T \subseteq [0, +\infty)$. The movement of the pollen grain will be therefore described by the family:

$$\{X_t\}_{t \in T} \tag{6.15}$$

of random variables. For all $\omega \in \Omega$, one obtains the *trajectory*:

$$t \rightsquigarrow X_t(\omega) \tag{6.16}$$

corresponding to a possible motion of the system. Moreover, in the above outlined formalism it makes sense to compute expressions of the form:

$$P\left(X_{t_1} \in A_1, \ldots, X_{t_n} \in A_n\right), \quad A_1, \ldots, A_n \in \mathcal{B}\left(\mathbb{R}^d\right) \tag{6.17}$$

As it is well known, the σ-field $\mathcal{F}$ contains all possible events, corresponding to all the possible statements one can formulate once an outcome of the experiment is registered; in the present context, however, the information obtained through an observation increases with time: at time t one has gained knowledge of the position of the particle for past times $s \leq t$, but not for future times $s > t$. This means that a key ingredient for the probabilistic description of the brownian motion is represented by the following definition:

Definition 6.1 We call **filtration** $\{\mathcal{F}_t\}_{t \in T}$, $T \subset \mathbb{R}^+$, a family of sub-σ-fields of $\mathcal{F}$ increasing with t, i.e. such that $\mathcal{F}_s \subset \mathcal{F}_t$ if $s \leq t$. We call **stochastic basis** a probability space endowed with a filtration, i.e. an object of the form:

$$\left(\Omega, \mathcal{F}, \{\mathcal{F}_t\}_{t \in T}, P\right) \tag{6.18}$$

Intuitively, $\mathcal{F}_t$ contains all events that it is possible to *discriminate*, that is to conclude whether have occurred or not, once the system has been observed up to the instant t. In other words, it represents all the information available up to time t and including time t.

We can now give the following general definition:

Definition 6.2 A **stochastic process** is an object of the form:

$$X = \left(\Omega, \mathcal{F}, \{\mathcal{F}_t\}_{t \in T}, \{X_t\}_{t \in T}, P\right) \tag{6.19}$$

where:

1. $\left(\Omega, \mathcal{F}, \{\mathcal{F}_t\}_{t \in T}, P\right)$ is a stochastic basis
2. $\{X_t\}_{t \in T}$ is a family of random variables taking values in a measurable space $(E, \mathcal{E})$:

$$X_t : \Omega \to E \tag{6.20}$$

and such that, for all t, X_t is $\mathcal{F}_t$-measurable. To express this circumstance, we say that $\{X_t\}_{t \in T}$ is **adapted** to the filtration.

The set E is called **state space** of the process: for the purpose of our applications, it will coincide with $\mathbb{R}^d$, endowed with the σ-field of Borel sets. Given $\omega \in \Omega$, the map:

$$T \ni t \rightsquigarrow X_t(\omega) \in E \tag{6.21}$$

is called a **trajectory** of the process, which is called (**almost certainly**) **continuous** if the set of points ω such that the corresponding trajectory is continuous has probability 1.

6.3.1 Finite-Dimensional Laws

Let $X = \left(\Omega, \mathcal{F}, \{\mathcal{F}_t\}_{t\in T}, \{X_t\}_{t\in T}, P \right)$ be a stochastic process and $\pi = (t_1, \dots, t_n)$ a finite set of instants in T, such that $t_1 < \cdots < t_n$. The map:

$$\Omega \ni \omega \rightsquigarrow \left(X_{t_1}(\omega), \dots, X_{t_n}(\omega) \right) \in E^n \qquad (6.22)$$

is clearly a random variable; let μ_π be its law:

$$\mu_\pi \left(A_1 \times \cdots \times A_n \right) = P \left(X_{t_1} \in A_1, \dots, X_{t_n} \in A_n \right), \quad A_1, \dots, A_n \in \mathcal{E}. \quad (6.23)$$

μ_π is called a **finite-dimensional law** of X. Two processes sharing the same finite-dimensional laws are said **equivalent**.

Given π, μ_π is a probability measure on $(E^n, \otimes^n \mathcal{E})$, $\otimes^n \mathcal{E}$ being the smallest σ-field containing all the sets of the form $A_1 \times \cdots \times A_n$, $A_j \in \mathcal{E}$. Finite-dimensional laws, by construction, satisfy a simple **consistency** property:

$$\mu_\pi \left(A_1 \times ..A_{i-1} \times E \times A_{i+1}.. \times A_n \right) = \mu_{\pi'} \left(A_1 \times ..A_{i-1} \times A_{i+1}.. \times A_n \right) \quad (6.24)$$

for all t and $\pi = (t_1, \dots, t_n)$, provided that $\pi' = (t_1, \dots t_{i-1}, t_{i+1}, \dots, t_n)$. On the contrary, we will say that a family of probability measures $\{\mu_\pi\}_\pi$ is **consistent** if it satisfies the property (6.24). We state without proof the following fundamental result:

Theorem 6.1 (Kolmogorov) *Let E be a complete and separable metric space, $\mathcal{E}$ the σ-field of Borel sets and $\{\mu_\pi\}_\pi$ a family of consistent probability measures. Let:*

$$\Omega \overset{def}{=} E^T = \{\omega : T \to E\} \qquad (6.25)$$

be the set of all functions ω from T to E, and:

$$\mathcal{F} \overset{def}{=} \mathcal{B}(E)^T \qquad (6.26)$$

the smallest σ-field containing the cylindrical sets:

$$\{\omega \in \Omega \mid \omega(t_1) \in A_1, \dots, \omega(t_n) \in A_n\} \qquad (6.27)$$

Let finally:

$$X_t(\omega) \overset{def}{=} \omega(t) \qquad (6.28)$$

and $\mathcal{F}_t = \sigma \left(X_s, s \le t \right)$, be the **natural filtration**.

Then there exists a **unique** *probability measure P on $(\Omega, \mathcal{F})$ such that the measures $\{\mu_\pi\}_\pi$ are the finite-dimensional laws of the stochastic process $X = \left(\Omega, \mathcal{F}, \{\mathcal{F}_t\}_{t\in T}, \{X_t\}_{t\in T}, P \right)$.*

Despite its complexity, the meaning of Kolmogorov's theorem is clear: for any family of consistent probability distributions there exists a unique stochastic process having precisely that probability distributions as finite-dimensional laws.

In many contexts, it is interesting to draw some conclusions about the regularity of the trajectories of a stochastic process. To this purpose, we preliminary mention that two stochastic processes $X = \left(\Omega, \mathcal{F}, \{\mathcal{F}_t\}_{t \in T}, \{X_t\}_{t \in T}, P\right)$ and $X' = \left(\Omega', \mathcal{F}', \{\mathcal{F}'_t\}_{t \in T}, \{X'_t\}_{t \in T}, P'\right)$ are called **modifications** of each other if $(\Omega, \mathcal{F}, P) = \left(\Omega', \mathcal{F}', P'\right)$ and if, for all $t \in T$, $X_t = X'_t$ almost surely.

Again we state without proof the following second Kolmogorov theorem:

Theorem 6.2 (Kolmogorov) *Let X be a process taking values in $\mathbb{R}^d$, and such that for suitable $\alpha > 0$, $\beta > 0$, $c > 0$ and for all instants s, t:*

$$E\left[|X_t - X_s|^\beta\right] \leq c|t - s|^{1+\alpha} \tag{6.29}$$

Then there exists a modification X' of X which is continuous.

6.4 Construction of Brownian Motion

We now have all the instruments for giving a precise definition of the brownian motion, starting from Einstein's discussion. We consider $\left(\mathbb{R}^d, \mathcal{B}\left(\mathbb{R}^d\right)\right)$ as state space of the process. We recall that in the historical description of the brownian motion a key role was played by the transition probability density, which turned out to be normal. Moreover, in Einstein's discussion, the assumption that the transition of the pollen grain is independent on its position at current and past time was implicit. Summing up these observations and completing the description of the brownian motion with a suitable initial condition we formulate the following:

Definition 6.3 A process $B = \left(\Omega, \mathcal{F}, \{\mathcal{F}_t\}_{t \geq 0}, \{B_t\}_{t \geq 0}, P\right)$ is called **brownian motion** if:

1. $B_0 = \mathbf{0}$ almost certainly
2. for all $0 \leq s \leq t$ the random variable $B_t - B_s$ is independent of B_u for all $u \leq s$, and in particular it is independent of $\mathcal{F}_s$;
3. for all $0 \leq s \leq t$ the random variable $B_t - B_s$ has law $N(\mathbf{0}, (t - s)\mathbb{I})$, where $\mathbb{I}$ is the $d \times d$ identity matrix.

The first property corresponds to the requirement that the particle starts its motion at the origin of a suitable reference frame. The second property is commonly resumed saying that **the increments of the brownian motion are independent of the past**, and is strongly related to the memoryless condition for Markov chains, as it will be soon explained in detail. The third property makes the introduction of a gaussian transition probability density rigorous.

As the reader might have observed, so far the stochastic basis of the brownian motion has not been defined. In fact, no arguments have been presented to guarantee the existence of a brownian motion: in the previous Definition 6.3 the requirements formulated in the heuristic introduction to the brownian motion have just been formulated in the language of stochastic processes.

To the purpose of proving the existence of the brownian motion, we first observe that the Definition 6.3 of brownian motion corresponds to assigning the finite-dimensional laws of the process. We will limit our treatment to the one-dimensional case $d = 1$, as the multidimensional case results from a straightforward generalization left to the reader.

First, we prove that the n-dimensional random variable $(B_{t_1}, \ldots, B_{t_n})$ is normal. By virtue of the second and third property of the brownian motion the vector random variable $(Y_{t_1}, \ldots, Y_{t_n})$ given by:

$$Y_{t_1} = B_{t_1} \quad Y_{t_k} = B_{t_k} - B_{t_{k-1}} \quad k = 2 \ldots n \tag{6.30}$$

has independent and normally distributed components with zero mean and covariance matrix:

$$\Delta = \begin{pmatrix} t_1 & 0 & \ldots & 0 \\ 0 & t_2 - t_1 & \ldots & 0 \\ \ldots & \ldots & \ldots & \ldots \\ 0 & 0 & \ldots & t_n - t_{n-1} \end{pmatrix} \tag{6.31}$$

Since it is related to the vector random variable $(B_{t_1}, \ldots, B_{t_n})$ by the linear transformation:

$$\begin{pmatrix} B_{t_1} \\ B_{t_2} \\ \ldots \\ B_{t_n} \end{pmatrix} = \begin{pmatrix} 1 & 0 & \ldots & 0 \\ 1 & 1 & \ldots & 0 \\ \ldots & \ldots & \ldots & \ldots \\ 1 & 1 & \ldots & 1 \end{pmatrix} \begin{pmatrix} Y_{t_1} \\ Y_{t_2} \\ \ldots \\ Y_{t_n} \end{pmatrix} = A \begin{pmatrix} Y_{t_1} \\ Y_{t_2} \\ \ldots \\ Y_{t_n} \end{pmatrix} \tag{6.32}$$

the former is also normally distributed with covariance matrix:

$$\Gamma = A \Delta A^T \tag{6.33}$$

Explicitly:

$$\Gamma_{ij} = E \left[B_{t_i} B_{t_j} \right] = \min(t_i, t_j) \tag{6.34}$$

This can be verified also with the following simple calculation, assuming, for example, $t_i < t_j$:

$$\begin{aligned} E \left[B_{t_i} B_{t_j} \right] &= E \left[B_{t_i} \left(B_{t_j} - B_{t_i} \right) \right] + E \left[B_{t_i}^2 \right] = \\ &= E \left[B_{t_i} \right] E \left[B_{t_j} - B_{t_i} \right] + t_i = t_i \end{aligned} \tag{6.35}$$

Since the covariance matrix Γ is invertible with inverse $\Gamma^{-1} = A^{-1} \Delta^{-1} \left(A^{-1} \right)^T$ where, as a simple calculation shows:

$$A^{-1} = \begin{pmatrix} 1 & 0 & 0 & \dots & 0 & 0 \\ -1 & 1 & 0 & \dots & 0 & 0 \\ \dots & \dots & \dots & \dots & \dots & \dots \\ 0 & 0 & 0 & \dots & -1 & 1 \end{pmatrix} \tag{6.36}$$

we can write the finite-dimensional laws of the brownian motion:

$$\mu_\pi (A_1 \times \dots \times A_n) = \int_{A_1} dx_1 \dots \int_{A_n} dx_n \frac{\exp\left(-\frac{1}{2}\left(\sum_{i,j=1}^n x_i \Gamma_{ij}^{-1} x_j\right)\right)}{(2\pi)^{n/2} \det(\Gamma)^{1/2}} \tag{6.37}$$

The above expression can be given a more transparent form since, due to Eqs. (6.33) and (6.36), the following simplifications occur:

$$\sum_{i,j=1}^n x_i \Gamma_{ij}^{-1} x_j = \frac{x_1^2}{t_1} + \sum_{k=2}^n \frac{(x_k - x_{k-1})^2}{(t_k - t_{k-1})} \tag{6.38}$$

$$\det(\Gamma) = t_1 (t_2 - t_1) \dots (t_n - t_{n-1})$$

leading to the identity:

$$\mu_\pi (A_1 \times \dots \times A_n) = \tag{6.39}$$

$$= \int_{A_1} dx_1 \, p(x_1, t_1|0, 0) \int_{A_2} dx_2 \, p(x_2, t_2|x_1, t_1) \dots \int_{A_n} dx_n p(x_n, t_n|x_{n-1}, t_{n-1})$$

where the *transition probability density* is given by:

$$p(y, t|x, s) = \frac{1}{\sqrt{2\pi(t-s)}} \exp\left(-\frac{(y-x)^2}{2(t-s)}\right), \quad s < t \tag{6.40}$$

In the multidimensional case one has:

$$p(\mathbf{y}, t|\mathbf{x}, s) = \frac{1}{[2\pi(t-s)]^{d/2}} \exp\left(-\frac{|\mathbf{y}-\mathbf{x}|^2}{2(t-s)}\right), \quad s < t \tag{6.41}$$

It is also immediate to realize that the expression (6.39) implies the three properties in the Definition 6.3 of the brownian motion. The first property is obvious. The second and third are retrieved choosing $n = 2$ and $A_1 = \mathbb{R}$ (Fig. 6.2):

$$\mu_\pi (A_2) = \int_{\mathbb{R}} dx_1 \, p(x_1, t_1|0, 0) \int_{A_2} dx_2 \, p(x_2, t_2|x_1, t_1) \tag{6.42}$$

which shows that $B_{t_2} = B_{t_1} + B_{t_2} - B_{t_1}$, with the increment $B_{t_2} - B_{t_1}$ independent on B_{t_1} and normally distributed with mean 0 and variance $t_2 - t_1$.

Fig. 6.2 Probability density
of B_t, $p(x, t)$, as a function
of the position and of time

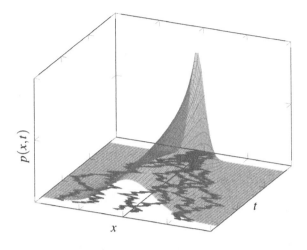

6.5   Transition Probability and Existence of the Brownian Motion

The Definition 6.3 of brownian motion given above is equivalent to the assignment of the finite-dimensional laws. The latter present an interesting structure, and involve an object closely recalling the transition matrices introduced in the context of Markov chains. Consider in fact the following map, called **markovian transition function of the brownian motion**:

$$p(A, t \,|\mathbf{x}, s) \stackrel{def}{=} \frac{1}{[2\pi (t - s)]^{d/2}} \int_A d\mathbf{y} \, \exp\left(-\frac{|\mathbf{y} - \mathbf{x}|^2}{2(t - s)}\right) \qquad (6.43)$$

where $s, t \in T$, $s < t$, $\mathbf{x} \in \mathbb{R}^d$ and $A \in \mathcal{B}\left(\mathbb{R}^d\right)$. If $s = t$, we put instead:

$$p(A, s \,|\mathbf{x}, s) \stackrel{def}{=} \delta_{\mathbf{x}}(A) \qquad (6.44)$$

Intuitively, we interpret $p(A, t \,|\mathbf{x}, s)$ as the probability that the pollen grain is in the Borel set A at time t, given the fact that it was at the point $\mathbf{x}$ at time s. Incidentally we observe that the probability of finding a particle at a precise point vanishes at any instant $s > 0$, so that the map (6.44) cannot be properly interpreted as conditional probability. On the other hand, we realize that, by definition, $B_t - B_0 \sim N(0, t)$, and therefore:

$$P(B_t \in A) = p(A, t \,|\mathbf{0}, 0) = \frac{1}{(2\pi t)^{d/2}} \int_A d\mathbf{y} \, \exp\left(-\frac{|\mathbf{y}|^2}{2t}\right) \qquad (6.45)$$

recalling that, by definition, $B_0 = \mathbf{0}$ almost surely. In general:

$$P(\mathbf{x} + B_t - B_s \in A) = p(A, t \mid \mathbf{x}, s) \tag{6.46}$$

We now introduce some observations, that will turn out to be useful later in the discussion:

1. for fixed s, t, A the function $\mathbf{x} \mapsto p(A, t \mid \mathbf{x}, s)$ is Borel-measurable
2. for fixed $s, t, \mathbf{x}$, the function $A \mapsto p(A, t \mid \mathbf{x}, s)$ is a probability measure on $\left(\mathbb{R}^d, \mathcal{B}\left(\mathbb{R}^d\right)\right)$;
3. p satisfies the following **Chapman-Kolmogorov equation**:

$$p(A, t \mid \mathbf{x}, s) = \int_{\mathbb{R}^d} p(A, t \mid \mathbf{y}, u) \, p(dy, u \mid \mathbf{x}, s) \tag{6.47}$$

for all $s < u < t, \mathbf{x} \in \mathbb{R}^d$ and $A \in \mathcal{B}\left(\mathbb{R}^d\right)$.

The reader is invited to derive the last Eq. (6.47) through a direct calculation, which is a simple exercise on gaussian integrals.

As proved earlier, the finite-dimensional laws of the brownian motion are given by:

$$\mu_\pi (A_1 \times \cdots \times A_n) = \int_{A_1} p(d\mathbf{x}_1, t_1 \mid \mathbf{0}, 0) \int_{A_2} \cdots \int_{A_n} p(d\mathbf{x}_n, t_n \mid \mathbf{x}_{n-1}, t_{n-1}) \tag{6.48}$$

where $\pi = (t_1, \ldots, t_n)$, $0 \le t_1 < \cdots < t_n$. The consistence of these finite-dimensional laws is an immediate consequence of the Chapman-Kolmogorov equation.

Kolmogorov's Theorem 6.1 ensures the **existence** of a stochastic process:

$$B = \left(\Omega, \mathcal{F}, \{\mathcal{F}_t\}_{t \ge 0}, \{B_t\}_{t \ge 0}, P\right) \tag{6.49}$$

having (6.48) as finite-dimensional laws, and makes it possible to construct the stochastic basis for such process. Ω is the set of trajectories:

$$\Omega = \{\omega : [0, +\infty) \to \mathbb{R}^d\} \tag{6.50}$$

$\mathcal{F}$ the σ-field $\mathcal{B}\left(\mathbb{R}^d\right)^{[0,+\infty)}$, appearing in Kolmogorov's theorem, and $\mathcal{F}_t = \sigma\left(X_s, s \le t\right)$ the natural filtration. The process is defined by:

$$\omega \in \Omega, \quad \omega \to B_t(\omega) \stackrel{def}{=} \omega(t) \tag{6.51}$$

The second Kolmogorov's theorem guarantees that a large class of stochastic processes can be modified in such a way as to be turned into a continuous process. To prove that the second Kolmogorov's theorem applies to the brownian motion, we consider $\beta = 2n$ for some integer $n \ge 2$ and $t > s$:

$$E\left[|B_t - B_s|^\beta\right] = \frac{1}{\sqrt{2\pi(t-s)}} \int_{\mathbb{R}} dx \, x^{2n} \exp\left(-\frac{x^2}{2(t-s)}\right) =$$

$$= \frac{2^n}{\sqrt{4\pi}} (t-s)^n \int_0^\infty du \, u^{n+\frac{1}{2}} \exp(-u) = \frac{2^n \, \Gamma\left(n+\frac{1}{2}\right)}{\sqrt{4\pi}} (t-s)^n = c \, (t-s)^{1+\alpha}$$

(6.52)

From now on, we will always assume to work with a continuous brownian motion.

We conclude this section with a technical observation. It is often useful, in particular to prove some theorems on stochastic calculus, to choose a filtration which is larger than the natural one, so that the stochastic basis:

$$\left(\Omega, \mathcal{F}, \{\mathcal{F}_t\}_{t\geq 0}, P\right)$$

(6.53)

satisfies the so-called **usual hypotheses**, that is:

1. the filtration is right continuous, i.e. $\mathcal{F}_t = \mathcal{F}_{t_+} \overset{def}{=} \bigcap_{s>t} \mathcal{F}_s$
2. each sub-σ-field $\mathcal{F}_t$ contains all the events of $\mathcal{F}$ with vanishing probability

A simple way to satisfy the usual hypotheses is add to all the σ-fields $\mathcal{F}_t = \sigma\left(X_s, s \leq t\right)$ all the events of $\mathcal{F}$ with vanishing probability. The so-obtained filtration is referred to as **completed natural filtration**.

6.6   The Martingale Property and the Markov Property

Due to the fact that the increments of the brownian motion are independent of the past, the stochastic process exhibits the following remarkable property:

$$E\left[B_t|\mathcal{F}_s\right] = E\left[B_t - B_s + B_s|\mathcal{F}_s\right] =$$
$$= E\left[B_t - B_s|\mathcal{F}_s\right] + E\left[B_s|\mathcal{F}_s\right] =$$
$$= E\left[B_t - B_s\right] + B_s = B_s$$

(6.54)

where the independence property and the fact that B_s is $\mathcal{F}_s$-measurable have been used. The above equation means that B_s is the best prediction for B_t given the information obtained observing the system up to time s. In the language of stochastic processes, this property is expressed saying that the brownian motion is a **martingale**.

The problem of determining $E\left[f(B_t)|\mathcal{F}_s\right]$, if $f : \mathbb{R}^d \to \mathbb{R}$ is a limited Borel function $f : \mathbb{R}^d \to \mathbb{R}$, has great interest. To this purpose, we will refer to a useful theorem we have presented in Chap. 3, and which we recall here for the sake of clarity:

Theorem 6.3 *Let $(\Omega, \mathcal{F}, P)$ be a probability space, $\mathcal{G}$ and $\mathcal{H}$ mutually independent sub-σ-fields of $\mathcal{F}$. Let $X : \Omega \to E$ be a $\mathcal{G}$-measurable random variable taking values* *in a measurable space* $(E, \mathcal{E})$ *and* ψ *a function* $\psi : E \times \Omega \to \mathbb{R}$ $\mathcal{E} \otimes \mathcal{H}$-*measurable and such that* $\omega \mapsto \psi(X(\omega), \omega)$ *is integrable. Then:*

$$E[\psi(X, \cdot)|\mathcal{G}] = \Phi(X), \quad \Phi(x) \stackrel{def}{=} E[\psi(x, \cdot)] \tag{6.55}$$

We know that $\mathcal{G} \equiv \mathcal{F}_s$ and $\mathcal{H} \equiv \sigma(B_t - B_s)$ are mutually independent, and that the random variable $X \equiv B_s$ is $\mathcal{G}$-measurable. Moreover the function $\psi : \mathbb{R}^d \times \Omega \to \mathbb{R}$ given by $\psi(\mathbf{x}, \omega) \equiv f(\mathbf{x} + B_t(\omega) - B_s(\omega))$ is $\mathcal{H}$-measurable. The function $\Phi(\mathbf{x})$ appearing in the above theorem is therefore:

$$\Phi(\mathbf{x}) = E[f(\mathbf{x} + B_t - B_s)] \tag{6.56}$$

and since the random variable $\omega \mapsto \mathbf{x} + B_t(\omega) - B_s(\omega)$ has law $N(\mathbf{x}, (t-s)\,\mathbb{I})$:

$$\Phi(\mathbf{x}) = \frac{1}{[2\pi(t-s)]^{d/2}} \int_{\mathbb{R}^d} d\mathbf{y}\, f(\mathbf{y}) \exp\left(-\frac{|\mathbf{y} - \mathbf{x}|^2}{2(t-s)}\right) \tag{6.57}$$

we conclude that:

$$E[f(B_t)|\mathcal{F}_s] = E[f(B_s + B_t - B_s)|\mathcal{F}_s] = E[\psi(B_s, \cdot)|\mathcal{F}_s] = \Phi(B_s) \tag{6.58}$$

and:

$$E[f(B_t)|\mathcal{F}_s] = \frac{1}{[2\pi(t-s)]^{d/2}} \int_{\mathbb{R}^d} d\mathbf{y}\, f(\mathbf{y}) \exp\left(-\frac{|\mathbf{y} - B_s|^2}{2(t-s)}\right) \tag{6.59}$$

This result is extremely important: the observation of the system up to time s corresponds to the knowledge of B_s, a circumstance already reported in Markov chains. In particular, if $f = 1_A$ for some $A \in \mathcal{B}(\mathbb{R}^d)$, the above expression becomes:

$$P(B_t \in A|\mathcal{F}_s) = p(A, t \mid B_s, s) \tag{6.60}$$

(6.60) is referred to as **Markov property** of the brownian motion.

6.7 Wiener Measure and Feynman Path Integral

Let's consider again the transition probability density:

$$p(\mathbf{y}, t \mid \mathbf{x}, s) = \frac{1}{[2\pi(t-s)]^{d/2}} \exp\left(-\frac{|\mathbf{y} - \mathbf{x}|^2}{2(t-s)}\right) \tag{6.61}$$

The Chapman-Kolmogorov property can be written in the form:

$$p\left(\mathbf{y}, t \mid \mathbf{x}, s\right) = \int_{\mathbb{R}^d} d\mathbf{x}_1 \, p\left(\mathbf{y}, t \mid \mathbf{x}_1, t_1\right) \, p\left(\mathbf{x}_1, t_1 \mid \mathbf{x}, s\right) \tag{6.62}$$

Suppose we iterate, introducing a partition $t_0 = s < t_1 < \cdots < t_n = t$ and letting $\mathbf{x}_0 = \mathbf{x}$, $\mathbf{x}_n = \mathbf{y}$. We get:

$$p\left(\mathbf{y}, t \mid \mathbf{x}, s\right) = \int \prod_{i=1}^{n-1} d\mathbf{x}_i \prod_{i=0}^{n-1} p\left(\mathbf{x}_{i+1}, t_{i+1} \mid \mathbf{x}_i, t_i\right) \tag{6.63}$$

that is:

$$p\left(\mathbf{y}, t \mid \mathbf{x}, s\right) = \int \prod_{i=1}^{n-1} d\mathbf{x}_i \prod_{i=0}^{n-1} \frac{1}{\left[2\pi\left(t_{i+1} - t_i\right)\right]^{d/2}} \exp\left(-\frac{|\mathbf{x}_{i+1} - \mathbf{x}_i|^2}{2(t_{i+1} - t_i)}\right) \tag{6.64}$$

We observe that the result is independent from the particular choice of the partition of the time interval. We can thus arbitrarily increase the number n, eventually making $n \to +\infty$. For example, if we choose a uniform partition, $t_{i+1} - t_i = \Delta t$, we observe that:

$$\prod_{i=0}^{n-1} \exp\left(-\frac{|\mathbf{x}_{i+1} - \mathbf{x}_i|^2}{2\Delta t}\right) = \exp\left(-\sum_{i=0}^{n-1} \frac{|\mathbf{x}_{i+1} - \mathbf{x}_i|^2}{2\Delta t^2} \Delta t\right) \tag{6.65}$$

and, for small Δt, we are induced to formally write:

$$\exp\left(-\sum_{i=0}^{n-1} \frac{|\mathbf{x}_{i+1} - \mathbf{x}_i|^2}{2\Delta t^2} \Delta t\right) \simeq \exp\left(-\int_s^t d\tau \frac{1}{2} \left|\frac{d\mathbf{x}(\tau)}{d\tau}\right|^2\right) \tag{6.66}$$

where the sequence of points $\mathbf{x}_i$ has become a *motion* $\mathbf{x}(\tau)$.

Motivated by this observation we **define** the **Path-Integral**:

$$\int_{\mathbf{x}(s)=\mathbf{x}, \, \mathbf{x}(t)=\mathbf{y}} \mathcal{D}\mathbf{x}(\tau) \, \exp\left(-\int_s^t d\tau \frac{1}{2} \left|\frac{d\mathbf{x}(\tau)}{d\tau}\right|^2\right) \tag{6.67}$$

as:

$$\lim_{t_{i+1}-t_i \to 0} \int \prod_{i=1}^{n-1} d\mathbf{x}_i \prod_{i=0}^{n-1} \frac{1}{\left[2\pi\left(t_{i+1} - t_i\right)\right]^{d/2}} \exp\left(-\frac{|\mathbf{x}_{i+1} - \mathbf{x}_i|^2}{2(t_{i+1} - t_i)}\right) \tag{6.68}$$

where the limit is meant by considering partitions of arbitrarily small width and arbitrarily large number of points.

The Chapman-Kolmogorov property guarantees that such definition is well posed and we know that:

$$p(\mathbf{y}, t \mid \mathbf{x}, s) = \int_{\mathbf{x}(s)=\mathbf{x},\, \mathbf{x}(t)=\mathbf{y}} \mathcal{D}\mathbf{x}(\tau)\, \exp\left(-\int_s^t d\tau\, \frac{1}{2}\left|\frac{d\mathbf{x}(\tau)}{d\tau}\right|^2\right) \qquad (6.69)$$

The interpretation of this identity is strongly related to Feynman approach to Quantum Mechanics: the transition probability for the brownian particle to move from $\mathbf{x}$ to $\mathbf{y}$ in the time interval $[s, t]$ is obtained by *summing* all the possible trajectories joining $\mathbf{x}$ and $\mathbf{y}$, each path being weighted by the exponential of the *free-particle action* functional (Fig. 6.3):

$$S[\mathbf{x}(\bullet)] = \int_s^t d\tau\, \frac{1}{2}\left|\frac{d\mathbf{x}(\tau)}{d\tau}\right|^2 \qquad (6.70)$$

We note that a classical limit is transparent: only the trajectory with minimal action contributes to the *summation*, making the free particle follow a straight line. The measure $\mathcal{D}\mathbf{x}(\tau)$, called the **Wiener measure**, is actually a well defined measure on the Wiener space $\mathcal{W}_d$ of continuous functions:

$$\mathbf{x} : [s, t] \to \mathbb{R}^d, \quad \tau \to \mathbf{x}(\tau) \qquad (6.71)$$

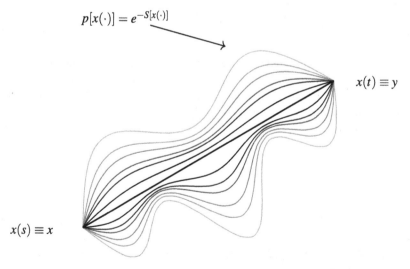

Fig. 6.3 Representation of possible trajectories joining x and y, each one being weighted by the exponential of the free-particle action functional

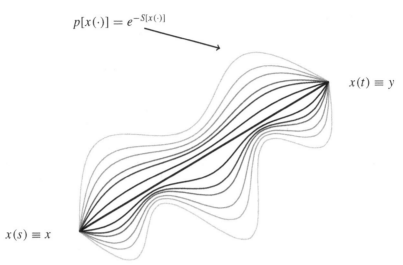

Historically, Feynman introduced this formalism of path integrals in quantum mechanics during the year 1948. His path integral uses real time, the weight being the imaginary number:

$$\exp\left(-\frac{i}{\hbar} S\left[\mathbf{x}(\bullet)\right]\right) \tag{6.72}$$

It is not possible to rigorously define the path integral of the above imaginary valued function, since the defining sequence does not converge in the limit of partitions of arbitrarily small width and arbitrarily large number of points. Nevertheless, this difficulty can however be overcome via a Wick rotation, $t \to \tau = -i\hbar t$, driving the quantum free particle into the brownian particle!

We observe that, in principle, it is very interesting to go beyond the free particle, introducing an action containing a potential energy: the stochastic motion will not be a brownian motion, but still a Markov process, as we will learn in the next chapter. We will come back to this point when we will have developed the foundations of the stochastic calculus.

6.8 Markov Processes

Moving from the discussion of the previous paragraphs, we will now abstract the general definition of Markov process, which extends to continuous time and state space the notion of Markov chain, introduced in the previous chapters. The discussion will begin with the following:

Definition 6.4 A **markovian transition function** on a measurable space $(E, \mathcal{E})$ is a real-valued function $p(A, t \mid x, s)$, where $s, t \in \mathbb{R}^+, s \leq t, x \in E$ and $A \in \mathcal{E}$ is such that:

1. for fixed s, t, A the function $x \mapsto p(A, t \mid x, s)$ is $\mathcal{E}$-measurable
2. for fixed s, t, x the function $A \mapsto p(A, t \mid x, s)$ is a probability measure on $(E, \mathcal{E})$

3. p satisfies the following **Chapman-Kolmogorov equation**:

$$p(A, t \mid x, s) = \int_E p(A, t \mid y, u) \, p(dy, u \mid x, s) \qquad (6.73)$$

for all $s < u < t, x \in E$ and $A \in \mathcal{E}$;

4. if $s = t, p(A, s \mid x, s) = \delta_x(A)$ for all $x \in E$.

The reader is invited to observe that p generalizes the n-step transition matrices of Markov chains.

Definition 6.5 Let $(E, \mathcal{E})$ be a measurable space. Given a markovian transition function p on $(E, \mathcal{E})$ and a probability law μ, we call **Markov process** associated to p, starting at time u with initial law μ, a process $X = \left(\Omega, \mathcal{F}, \{\mathcal{F}_t\}_{t \in T}, \{X_t\}_{t \in T}, P\right)$, with time domain $T = [u, +\infty)$ and state space $(E, \mathcal{E})$, such that:

1. X_u has law μ;
2. the following **Markov property** holds:

$$P(X_t \in A \mid \mathcal{F}_s) = p(A, t \mid X_s, s) \qquad (6.74)$$

almost surely for all $A \in \mathcal{E}$ and $t > s \geq u$.

The brownian motion is our first example of Markov process.

A fascinating and surprising relationship between theory of stochastic processes and theory of partial differential equation exists. In the remainder of the present chapter a first discussion providing an insight into this topic will be raised, and deepened once the formalism of stochastic differential equations will have been introduced.

6.9 Semigroup Associated to a Markovian Transition Function

Let p be a markovian transition function on $(\mathbb{R}^d, \mathcal{B}(\mathbb{R}^d))$, which for the moment will be assumed **time homogeneous**, i.e. depending only on $t - s$. If f is a real-valued and bounded Borel function, we can define the time dependent family of operators:

$$(T_t f)(\mathbf{x}) \overset{def}{=} \int_{\mathbb{R}^d} d\mathbf{y}\, f(\mathbf{y})\, p(d\mathbf{y}, t \mid \mathbf{x}, 0) \tag{6.75}$$

Intuitively this operator, which we will meet again in the future chapters, represents a time dependent average of a given function of a Markov process, for example the energy of a system which evolves in time under the action of random external fields. We immediately see that T_0 is the identity, and that the following composition property holds:

$$T_s \circ T_t = T_{s+t} \tag{6.76}$$

To prove (6.76), let us compute:

$$\begin{aligned}
((T_s \circ T_t)f)\mathbf{x}) &= \int_{\mathbb{R}^d} d\mathbf{y}\, (T_t f)(\mathbf{y})\, p(d\mathbf{y}, s \mid \mathbf{x}, 0) = \qquad (6.77)\\
&= \int_{\mathbb{R}^d} d\mathbf{y} \int_{\mathbb{R}^d} d\mathbf{z}\, f(\mathbf{z}) p(d\mathbf{z}, t \mid \mathbf{y}, 0) p(d\mathbf{y}, s \mid \mathbf{x}, 0) = \\
&= \int_{\mathbb{R}^d} d\mathbf{z}\, f(\mathbf{z}) \int_{\mathbb{R}^d} d\mathbf{y} p(d\mathbf{z}, s+t \mid \mathbf{y}, s)\, p(d\mathbf{y}, s \mid \mathbf{x}, 0) = \\
&= \int_{\mathbb{R}^d} d\mathbf{z}\, f(\mathbf{z})\, p(d\mathbf{z}, s+t \mid \mathbf{x}, 0) = (T_{s+t} f)(\mathbf{x})
\end{aligned}$$

where the homogeneity of the markovian transition function and the Chapman-Kolmogorov equation have been recalled. Hence $\{T_t\}_{t \geq 0}$ is a **semigroup** of linear operators acting on the space of real-valued and bounded Borel functions.

If, for all $\mathbf{x} \in \mathbb{R}^d$, the following limit exists:

$$\lim_{t \to 0^+} \frac{1}{t} [(T_t f)(\mathbf{x}) - f(\mathbf{x})] \tag{6.78}$$

we can define the **infinitesimal generator** $\mathcal{L}$ of the semigroup $\{T_t\}_{t \geq 0}$:

$$(\mathcal{L}f)(\mathbf{x}) \overset{def}{=} \lim_{t \to 0^+} \frac{1}{t} [(T_t f)(\mathbf{x}) - f(\mathbf{x})] \tag{6.79}$$

which, intuitively, is the time derivative, at $t = 0$, of the time dependent average.

More generally, if the markovian transition function is **not time homogeneous**, the same reasoning leads to a family of operators $\{T_{s,t}\}_{s \leq t}$, defined through:

$$(T_{s,t} f)(\mathbf{x}) \overset{def}{=} \int_{\mathbb{R}^d} d\mathbf{y}\, f(\mathbf{y})\, p(d\mathbf{y}, t \mid \mathbf{x}, s) \tag{6.80}$$

$T_{s,s}$ is the identity, and:

$$T_{s,u} \circ T_{u,t} = T_{s,t}, \quad s \leq u \leq t \tag{6.81}$$

Instead of the operator $\mathcal{L}$ a family of infinitesimal generators $\{\mathcal{L}_t\}_t$ appears, defined through the expression:

$$(\mathcal{L}_t f)(\mathbf{x}) \overset{def}{=} \lim_{h \to 0^+} \frac{1}{h} \left[(T_{t,t+h} f)(\mathbf{x}) - f(\mathbf{x}) \right] \tag{6.82}$$

whenever it makes sense.

If f is smooth, more precisely if $f \in C^2(\mathbb{R}^d)$, the explicit expression:

$$\frac{1}{h} \left[(T_{t,t+h} f)(\mathbf{x}) - f(\mathbf{x}) \right] = \frac{1}{h} \int_{\mathbb{R}^d} d\mathbf{y} \, (f(\mathbf{y}) - f(\mathbf{x})) \, p(d\mathbf{y}, t + h \mid \mathbf{x}, t) \tag{6.83}$$

can be combined with a Taylor expansion:

$$f(\mathbf{y}) - f(\mathbf{x}) = \sum_\alpha \frac{\partial f(\mathbf{x})}{\partial x_\alpha} (y_\alpha - x_\alpha) + \frac{1}{2} \sum_{\alpha,\beta} \frac{\partial^2 f(\mathbf{x})}{\partial x_\alpha \partial x_\beta} (y_\alpha - x_\alpha)(y_\beta - x_\beta) + o(|\mathbf{y} - \mathbf{x}|^2) \tag{6.84}$$

which immediately suggests a very interesting relation:

$$(\mathcal{L}_t f)(\mathbf{x}) = \frac{1}{2} \sum_{\alpha,\beta=1}^d a_{\alpha\beta}(\mathbf{x}, t) \frac{\partial^2 f}{\partial x_\alpha \partial x_\beta}(\mathbf{x}) + \sum_{\alpha=1}^d b_\alpha(\mathbf{x}, t) \frac{\partial f}{\partial x_\alpha}(\mathbf{x}) \tag{6.85}$$

with $a_{\alpha\beta}(\mathbf{x}, t)$ and $b_\alpha(\mathbf{x}, t)$ uniquely determined by the transition probability p:

$$a_{\alpha\beta}(\mathbf{x}, t) = \lim_{h \to 0^+} \frac{1}{h} \int_{\mathbb{R}^d} d\mathbf{y} \, (y_\alpha - x_\alpha)(y_\beta - x_\beta) \, p(d\mathbf{y}, t + h \mid \mathbf{x}, t)$$
$$b_\alpha(\mathbf{x}, t) = \lim_{h \to 0^+} \frac{1}{h} \int_{\mathbb{R}^d} d\mathbf{y} \, (y_\alpha - x_\alpha) \, p(d\mathbf{y}, t + h \mid \mathbf{x}, t) \tag{6.86}$$

provided that the two limits make sense. The expression (6.85) can be shown [3] to hold under suitable hypotheses involving the behavior of the transition probability p for small h, which has to ensure us that higher order terms in the Taylor expansion can be neglected. For completeness, we mention also that the matrix $\{a_{\alpha\beta}(\mathbf{x}, t)\}_{\alpha\beta}$ can be shown to be always positive-semidefinite [3].

The relation (6.85) is very important since in provides a bridge between two apparently independent branches of Mathematics: the theory of partial differential equations and the theory of stochastic processes. We will come back to this crucial point in the following chapters.

6.10   The Heat Equation

As shown in the previous paragraph, under suitable condition, to each Markov process, or equivalently to each markovian transition function, a differential operator can be associated. In the time-homogeneous case, which we consider only for the sake of simplicity, we know that the function $u(\mathbf{x}, t) = T_t f(\mathbf{x})$ satisfies:

$$\frac{\partial u}{\partial t}(\mathbf{x}, 0) = \lim_{h \to 0} \frac{1}{h}(T_h f(\mathbf{x}) - f(\mathbf{x})) = \mathcal{L} f(\mathbf{x}) \tag{6.87}$$

and that:

$$\frac{\partial u}{\partial t}(\mathbf{x}, t) = \lim_{h \to 0} \frac{1}{h}(T_{t+h} f(\mathbf{x}) - T_t f(\mathbf{x})) = T_t \mathcal{L} f(\mathbf{x}) \tag{6.88}$$

If it happens that T_t and $\mathcal{L}$ commute, the following **heat equation** associated to the markovian transition function is found:

$$\begin{cases} \frac{\partial u}{\partial t}(\mathbf{x}, t) = (\mathcal{L} u)(\mathbf{x}, t) \\ u(\mathbf{x}, 0) = f(\mathbf{x}) \end{cases} \tag{6.89}$$

On the other hand, let $q(t, \mathbf{x}, \mathbf{y})$ be the *fundamental solution* of the heat equation, i.e. the function satisfying:

$$\begin{cases} \frac{\partial q}{\partial t}(t, \mathbf{x}, \mathbf{y}) = (\mathcal{L} q)(t, \mathbf{x}, \mathbf{y}) \\ q(0, \mathbf{x}, \mathbf{y}) = \delta(\mathbf{x} - \mathbf{y}) \end{cases} \tag{6.90}$$

then, from the theory of partial differential equations, it is known that:

$$u(\mathbf{x}, t) = \int_{\mathbb{R}^d} d\mathbf{y}\, q(t, \mathbf{x}, \mathbf{y}) f(\mathbf{y}) \tag{6.91}$$

But since $u(\mathbf{x}, t) = T_t f(\mathbf{x})$, the markovian transition function must be:

$$p(d\mathbf{y}, s + t \,|\, \mathbf{x}, s) = q(t, \mathbf{x}, \mathbf{y})\, d\mathbf{y} \tag{6.92}$$

This observation sheds light on the possibility of *inverting* this process: given a partial differential equation with fundamental solution q, we might ask ourselves whether it is possible to construct a Markov process having a markovian transition function p related to q by (6.92). As pointed out before, for this issue to be faced adequately, the formalism of stochastic differential equations is required.

6.11 Further Readings

Brownian motion is the cornerstone of the theory of stochastic processes and is presented in many excellent books. The presentation in this chapter is self-contained. A few books that readers may see to deepen their study are [1–3]. Readers interested in Feynman path integral can see the original book [4]. For the connection with the theory of partial differential equations we refer to the bibliography of the following chapters.

Problems

6.1 Invariances of the brownian motion

Given a one-dimensional brownian motion continuous with increments independent from the past B_t, show that $X_t = -B_t$ and $Y_t = \frac{1}{\sqrt{u}} B_{ut}$ are brownian motions.

6.2 Drifted brownian motion

Given a one-dimensional brownian motion continuous with increments independent from the past B_t, consider the process $X_t = B_t + vt, v \in \mathbb{R}$. Compute $m(t) = E[X_t]$, the probability density of X_t, and $C(t, s) = E[X_t X_s]$.

6.3 Distance from the origin

Given two independent brownian motions continuous with increments independent from the past $B_{1,t}$ and $B_{2,t}$, let:

$$R_t = \sqrt{B_{1,t}^2 + B_{2,t}^2} \tag{6.93}$$

Compute the probability density, the mean and the variance of R_t.

6.4 Brownian motion on the unit circle

Given a one-dimensional brownian motion continuous with increments independent from the past B_t, the stochastic process:

$$R_t = \begin{pmatrix} \cos(B_t) \\ \sin(B_t) \end{pmatrix} \tag{6.94}$$

is called brownian motion on the unit circle. Find the law of the random variables $\cos(B_t)$, $\sin(B_t)$ and $B_t \bmod 2\pi$. The first two random variables correspond to the projections of R_t onto the x and y axis, and the latter to the angle of the particle

6.5 Brownian bridge

Given a one-dimensional brownian motion continuous with increments independent from the past B_t, define the brownian bridge:

$$X_t = B_t - t B_1, \quad t \in [0, 1] \tag{6.95}$$

Show that X_t is normal, compute mean and variance. Show also that $(X_{t_1}, \ldots, X_{t_n})$ is normal. Compute mean and covariance matrix.

6.6 Correlated brownian motion

Given two independent brownian motions continuous with increments independent from the past $B_{1,t}$ and $B_{2,t}$, let:

$$X_t = \rho B_{1,t} + \sqrt{1 - \rho^2}\, B_{2,t} \tag{6.96}$$

Show that X_t is a brownian motion. Evaluate $E[X_t B_{1,t}]$.

6.7 Brownian bomb

Given a one-dimensional brownian motion continuous with increments independent from the past B_t and a random variable T exponential with parameter λ, find the probability density of the random variable $\omega \rightarrow Z(\omega) = B_{T(\omega)}(\omega)$.

References

1. Karatzas, I., Shreve, S.E.: Brownian Motion and Stochastic Calculus. Springer (2005)
2. Capasso, V., Bakstein, D.: An Introduction to Continuous Time Stochastic Processes. Birkhäuser (2005)
3. Baldi, P.: Equazioni Differenziali Stocastiche e Applicazioni. Pitagora Editori (2000)
4. Feynman, R.P., Hibbs, A.R.: Quantum Mechanics and Path Integrals. McGraw-Hill (1965)

Chapter 7
Stochastic Calculus and Introduction to Stochastic Differential Equations

Abstract In this chapter we introduce the basic notions of stochastic calculus, starting from the Brownian motion. Stochastic processes describe time-dependent random phenomena, generalizing the usual deterministic evolution. The description of the latter requires the notions of differential and integral, which need to be properly extended to stochastic properties. Stochastic calculus is the branch of mathematics dealing with this important topic. The reason why traditional calculus is not suitable for stochastic processes is revealed by the Brownian motion. Since $Var(B_t) = t$, implying that B_t "scales" as $\sqrt{t}$, its trajectories are not differentiable in the usual sense. The stochastic calculus allows us to introduce generalized notions of differential and integral, notwithstanding this difficulty. Moreover, it allows to write differential equations involving stochastic processes, providing thus a powerful generalization of ordinary differential equations to study phenomena evolving in time in a non deterministic way.

Keywords Stochastic calculus · Itô integral · Wiener integral
Approach to equilibrium · Langevin equation

7.1 Introduction

In this chapter we pursue the exploration of stochastic processes begun in Chap. 6 with the study of the Brownian motion. In order to deal with stochastic processes, which describe random phenomena depending on time, we need to introduce several basic notions and mathematical instruments. To understand the reason of this need, let us consider a system whose state at time t is described by a set of variables $\mathbf{x}_t$. Let us suppose that $\mathbf{x}_t$ undergoes a deterministic time evolution, i.e. an evolution in which no randomness is involved. This evolution, in most situations, is governed by an ordinary differential equation of the form:

$$\begin{cases} d\mathbf{x}_t = \mathbf{F}(\mathbf{x}_t, t)\, dt \\ \quad \mathbf{x}_0 = \mathbf{x}^* \end{cases} \tag{7.1}$$

© Springer International Publishing AG, part of Springer Nature 2018
E. Vitali et al., *Theory and Simulation of Random Phenomena*, UNITEXT
for Physics, https://doi.org/10.1007/978-3-319-90515-0_7

whose formal solution is readily obtained integrating both sides of (7.1) with respect to time:

$$\mathbf{x}_t = \mathbf{x}^* + \int_0^t \mathbf{F}\,(\mathbf{x}_s, s)\;ds \tag{7.2}$$

The solution of (7.2) defines a family $\{\mathbf{x}_t\}_t$ of states of the system. In the realm of stochastic processes, this family is replaced by a family of random variables $\{X_t\}_t$. The time evolution of the family $\{X_t\}_t$ is now governed by a stochastic differential equation: the latter should include a term introducing randomness in the time evolution.

This observation suggests that we should be able to rigorously define a notion of differential for a stochastic process $\{X_t\}_t$. Traditional calculus is not suitable for stochastic processes. The reason lies in the behavior of the trajectories of the Brownian motion: informally speaking, since $Var(B_t) = t$ the Brownian motion scales as $\sqrt{t}$ and thus has non-differentiable trajectories.

Equivalently, we should be able to rigorously define a notion of integral generalising (7.2) to stochastic processes: this crucial instrument is called the Itô integral after the work of Kiyoshi Itô. It is a generalisation of the ordinary concept of Riemann integral, which takes into account two important circumstances: (i) the integrand is a random variable and not an ordinary function and (ii) if randomness is present in the time evolution, integration is not only performed with respect to time but also with respect to the Brownian motion.

In order to properly fix the mathematical framework, for the time being we assign a stochastic basis in the **usual hypothesis** (defined in Sect. 6.5 in Chap. 6):

$$\left(\Omega, \mathcal{F}, \{\mathcal{F}_t\}_{t\geq 0}, P\right) \tag{7.3}$$

where we suppose a continuous Brownian motion is defined, with increments independent of the past. For simplicity, we now consider only the one-dimensional case:

$$B = \left(\Omega, \mathcal{F}, \{\mathcal{F}_t\}_{t\geq 0}, \{B_t\}_{t\geq 0}, P\right) \tag{7.4}$$

Remark 7.1 Having fixed the stochastic basis, we will use the simple notation $X = \{X_t\}_{t\geq 0}$ to indicate processes. Naturally, we will be careful to verify that all processes are adapted, i.e. that X_t is $\mathcal{F}_t$-measurable.

We start from the observation that a stochastic process $X = \{X_t\}_{t\geq 0}$ can be viewed as a function of two variables:

$$X : [0, +\infty) \times \Omega \to \mathbb{R}, \quad (t, \omega) \to X_t(\omega) \tag{7.5}$$

where each ω determines a *random trajectory* $t \to X_t(\omega)$ and t is a time instant.

We will deal with stochastic processes whose time evolution is governed by the
following integral equation, generalizing the deterministic one (7.2):

$$X_t = X_0 + \int_0^t F\,(X_s, s)\,ds + \int_0^t G\,(X_s, s)\,dB_s \qquad (7.6)$$

where the last term, introducing randomness, is constructed from the Brownian
motion.

In this chapter, we will learn to define the two kinds of integration of stochastic
processes appearing in (7.6), one with respect to time, and the other with respect to
the Brownian motion:

$$\int_\alpha^\beta F_s\,ds, \quad \int_\alpha^\beta G_s\,dB_s \qquad (7.7)$$

and to give a precise meaning to expressions of the form:

$$dX_t = F_t dt + G_t dB_t \qquad (7.8)$$

which will turn out to be a very important tool to build new processes starting from the
Brownian motion and will be the cornerstone of the theory of stochastic differential
equations.

In this chapter we will provide rigorous definitions and results about stochastic calculus, together with examples and applications. We will omit the proofs of
several theorems, which would require more advanced tools and can be found in
excellent textbooks on this subject. The Fig. 7.1 provides an illustration of the difference between integration with respect to time and integration with respect to the
brownian motion.

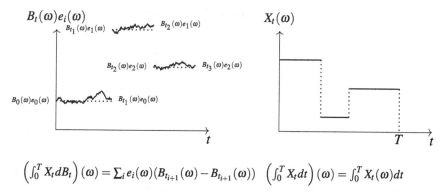

$$\left(\int_0^T X_t dB_t\right)(\omega) = \sum_i e_i(\omega)(B_{t_{i+1}}(\omega) - B_{t_{i+1}}(\omega)) \quad \left(\int_0^T X_t dt\right)(\omega) = \int_0^T X_t(\omega)dt$$

Fig. 7.1 Illustration of the two kinds of integration we will define in this chapter

7.2 Integration of Processes with Respect to Time

Let us consider a process X, interpreted as a function of two variables:

$$X : [0, +\infty) \times \Omega \to \mathbb{R}, \quad (t, \omega) \to X_t(\omega) \tag{7.9}$$

From the very definition of stochastic process we know that, for fixed t, the function: $\omega \to X_t(\omega)$ is $\mathcal{F}$-measurable and in particular $\mathcal{F}_t$-measurable. On the other hand, for fixed ω, the real-valued function $t \to X_t(\omega)$ defined on $[0, +\infty)$ has no assumed measurability properties.

We will focus our attention to **progressively measurable** processes, that is, by definition, processes such that, for all $\bar{t} > 0$, the function $(t, \omega) \to X_t(\omega)$ is $\left(\mathcal{B}([0, \bar{t}]) \otimes \mathcal{F}_{\bar{t}}\right)$-measurable. It is possible to show [1] that such technical hypothesis holds if the given process is continuous, i.e. it has continuous trajectories.

For such processes, the function:

$$t \to X_t(\omega) \tag{7.10}$$

is measurable and thus we can define its Lebesgue integral $\int_\alpha^\beta X_s ds$, one for each trajectory, which can be finite, infinite or even non existent:

$$\left(\int_\alpha^\beta X_s ds \right)(\omega) \overset{def}{=} \int_\alpha^\beta X_s(\omega) ds \tag{7.11}$$

where we have fixed a time interval $[\alpha, \beta], 0 \leq \alpha < \beta < +\infty$.

If the integral in (7.11) exists and is finite, owing to the hypothesis of progressive measurability, it is a random variable. It is thus natural to introduce the following:

Definition 7.1 Let X and Y two **progressively measurable** processes such that:

$$P\left(\int_\alpha^\beta |X_s| ds < +\infty \right) = 1 \tag{7.12}$$

We say that X and Y are **equivalent** if:

$$P\left(\int_\alpha^\beta |X_t - Y_t| dt = 0 \right) = 1 \tag{7.13}$$

We call $\Lambda^1(\alpha, \beta)$ the set made of equivalence classes of **progressively measurable** processes satisfying (7.12), the equivalence relation being defined by (7.13).

As usual, often the difference between a process and an equivalence class of processes is neglected.

Since we have chosen to work under the usual hypothesis, in every equivalence class in $\Lambda^1(\alpha, \beta)$ we can always find a representative X such that the integral:

$$\left(\int_\alpha^\beta X_s ds \right)(\omega) \tag{7.14}$$

exists for any ω. Such integral defines a real random variable $\mathcal{F}_\beta$-measurable and, if:

$$\int_\alpha^\beta E\left[|X_s|\right] ds < +\infty \tag{7.15}$$

then, by the classical Fubini's [2] theorem:

$$E\left[\int_\alpha^\beta X_s ds \right] = \int_\alpha^\beta E\left[X_s\right] ds \tag{7.16}$$

Moreover, if $X \in \Lambda^1(0, T)$, then the function:

$$(t, \omega) \to \int_0^t X_s ds, \quad t \in [0, T] \tag{7.17}$$

defines a stochastic process:

$$\left\{ \int_0^t X_s ds \right\}_t \tag{7.18}$$

which is **continuous**, since any integral is continuous with respect to the extremum, and thus **progressively measurable**.

To summarize, a stochastic process, under some quite natural hypothesis, can be integrated with respect to time: this is a simple Lebesgue integral of the single trajectories. In order to familiarize with this notion, we invite the reader to solve Problem 7.1.

7.3 The Itô Integral

We have defined the integral of a stochastic process with respect to time. We now define a different kind of integration, with respect to the Brownian motion. The reader could find useful to review the definition of the abstract integral with respect to the probability measure in the first chapter, Sect. 1.5, as the two constructions share some common features: the definition of Itô integral is first given within a particular class of stochastic processes, and then extended to a wider class of processes that cover most of the practical applications.

7.3.1 Itô Integral Integral of Simple Processes

Let us fix, as before, a time interval $[\alpha, \beta]$, $0 \le \alpha < \beta < +\infty$ and start with the following:

Definition 7.2 We say that a stochastic process X is **simple** if:

$$X_t(\omega) = \sum_{i=0}^{n-1} e_i(\omega) 1_{[t_i, t_{i+1})}(t) + e_n(\omega) 1_{\{\beta\}}(t) \tag{7.19}$$

for some choice of the integer n and of the times $\alpha = t_0 < \cdots < t_n = \beta$. The random variables e_i are $\mathcal{F}_{t_i}$-**measurable**.

For every ω, an simple process remains equal to a constant $e_i(\omega)$ over finite intervals of time $[t_i, t_{i+1})$. By construction, such a process is **progressively measurable**. We give now the first basic:

Definition 7.3 If X is an simple process, we call **Itô integral** or **stochastic integral** of X, and we denote it $\int_\alpha^\beta X_s d B_s$, the random variable:

$$\left(\int_\alpha^\beta X_s d B_s \right)(\omega) \overset{def}{=} \sum_{i=0}^{n-1} e_i(\omega) \left(B_{t_{i+1}}(\omega) - B_{t_i}(\omega) \right) \tag{7.20}$$

We denote $\mathcal{S}(\alpha, \beta)$ the set of simple processes, and $\mathcal{S}^2(\alpha, \beta)$ the set of **square-integrable** simple processes, i.e. $E\left[|X_t|^2 \right] < +\infty$.
   Naturally, $X \in \mathcal{S}^2(\alpha, \beta)$ if and only if $E[e_i^2] < +\infty$ for all i. Let us study now the map:

$$X \in \mathcal{S}(\alpha, \beta), \quad X \to I(X) \overset{def}{=} \int_\alpha^\beta X_s d B_s \tag{7.21}$$

in more detail. It has some important properties:

1. it is **linear**;
2. since the Brownian motion is adapted, $I(X)$ is $\mathcal{F}_\beta$-**measurable** because it depends only of random variables at times preceding β;
3. it satisfies the following additivity property:

$$\int_\alpha^\beta X_s d B_s = \int_\alpha^\gamma X_s d B_s + \int_\gamma^\beta X_s d B_s, \quad \alpha < \gamma < \beta \tag{7.22}$$

4. If $X \in \mathcal{S}^2(\alpha, \beta)$, then $I(X)$ is square integrable:

$$E\left[\left(\int_\alpha^\beta X_s d B_s \right)^2 \right] < +\infty \tag{7.23}$$

To verify this property, we have to show that $e_i e_j \left(B_{t_{i+1}} - B_{t_i} \right) \left(B_{t_{j+1}} - B_{t_j} \right)$ is integrable. If $i = j$, we know that e_i^2 is integrable since $X \in \mathcal{S}^2(\alpha, \beta)$; $\left(B_{t_{i+1}} - B_{t_i} \right)^2$ is naturally integrable (since it is a Brownian motion) and independent of e_i^2, which is $\mathcal{F}_{t_i}$-measurable. Thus $e_i^2 \left(B_{t_{i+1}} - B_{t_i} \right)^2$ is integrable, being the product of integrable independent random variables. This implies that $e_i \left(B_{t_{i+1}} - B_{t_i} \right)$ is square-integrable, and thus $e_i e_j \left(B_{t_{i+1}} - B_{t_i} \right) \left(B_{t_{j+1}} - B_{t_j} \right)$ is integrable, being a product of square-integrable random variables.

We have thus defined a linear map:

$$I : \mathcal{S}^2(\alpha, \beta) \to L^2(\Omega, \mathcal{F}_\beta, P), \quad X \to I(X) \stackrel{def}{=} \int_\alpha^\beta X_s dB_s \qquad (7.24)$$

Since $I(X)$ is square-integrable, it is also integrable because the probability P is a finite measure. The expectation $E[I(X)]$ is thus well defined and can be computed as follows:

$$E\left[\int_\alpha^\beta X_s dB_s | \mathcal{F}_\alpha \right] = E\left[\sum_{i=0}^{n-1} e_i \left(B_{t_{i+1}} - B_{t_i} \right) | \mathcal{F}_\alpha \right] =$$
$$= \sum_{i=0}^{n-1} E\left[E\left[e_i \left(B_{t_{i+1}} - B_{t_i} \right) | \mathcal{F}_{t_i} \right] | \mathcal{F}_\alpha \right] = \sum_{i=0}^{n-1} E\left[e_i E\left[\left(B_{t_{i+1}} - B_{t_i} \right) | \mathcal{F}_{t_i} \right] | \mathcal{F}_\alpha \right] =$$
$$= \sum_{i=0}^{n-1} E\left[e_i E\left[\left(B_{t_{i+1}} - B_{t_i} \right) \right] | \mathcal{F}_\alpha \right] = 0 \qquad (7.25)$$

where we have used the fact that e_i is $\mathcal{F}_{t_i}$-measurable and that $\left(B_{t_{i+1}} - B_{t_i} \right)$ is independent of $\mathcal{F}_{t_i}$, together with the properties of conditional expectation detailed in Sect. 3.6.

Thus we have found that:

$$E\left[\int_\alpha^\beta X_s dB_s | \mathcal{F}_\alpha \right] = 0 \qquad (7.26)$$

which implies, taking the expectation of both members, that:

$$E\left[\int_\alpha^\beta X_s dB_s \right] = 0 \qquad (7.27)$$

Let us now evaluate:

$$E\left[\left(\int_\alpha^\beta X_s dB_s \right)^2 | \mathcal{F}_\alpha \right] =$$
$$= \sum_i E\left[e_i^2 \left(B_{t_{i+1}} - B_{t_i} \right)^2 | \mathcal{F}_\alpha \right] + \sum_{i \neq j} E\left[e_i e_j \left(B_{t_{i+1}} - B_{t_i} \right) \left(B_{t_{j+1}} - B_{t_j} \right) | \mathcal{F}_\alpha \right] \qquad (7.28)$$

If $i < j$, we have:

$$E\left[e_i e_j \left(B_{t_{i+1}} - B_{t_i}\right)\left(B_{t_{j+1}} - B_{t_j}\right) | \mathfrak{F}_\alpha\right] =$$
$$= E\left[E\left[e_i e_j \left(B_{t_{i+1}} - B_{t_i}\right)\left(B_{t_{j+1}} - B_{t_j}\right) | \mathfrak{F}_{t_j}\right] | \mathfrak{F}_\alpha\right] =$$
$$= E\left[e_i e_j \left(B_{t_{i+1}} - B_{t_i}\right) E\left[\left(B_{t_{j+1}} - B_{t_j}\right) | \mathfrak{F}_{t_j}\right] | \mathfrak{F}_\alpha\right] = 0 \qquad (7.29)$$

so that no contribution arises from non-diagonal terms of (7.28). We have thus:

$$E\left[\left(\int_\alpha^\beta X_s dB_s\right)^2 | \mathfrak{F}_\alpha\right] =$$
$$= \sum_i E\left[e_i^2 \left(B_{t_{i+1}} - B_{t_i}\right)^2 | \mathfrak{F}_\alpha\right] =$$
$$= \sum_i E\left[E\left[e_i^2 \left(B_{t_{i+1}} - B_{t_i}\right)^2 | \mathfrak{F}_{t_i}\right] | \mathfrak{F}_\alpha\right] =$$
$$= \sum_i E\left[e_i^2 E\left[\left(B_{t_{i+1}} - B_{t_i}\right)^2 | \mathfrak{F}_{t_i}\right] | \mathfrak{F}_\alpha\right] =$$
$$= \sum_i E\left[e_i^2 E\left[\left(B_{t_{i+1}} - B_{t_i}\right)^2\right] | \mathfrak{F}_\alpha\right] =$$
$$= E\left[\sum_i e_i^2 (t_{i+1} - t_i) | \mathfrak{F}_\alpha\right] = E\left[\int_\alpha^\beta X_s^2 ds | \mathfrak{F}_\alpha\right] \qquad (7.30)$$

We conclude that:

$$E\left[\left(\int_\alpha^\beta X_s dB_s\right)^2 | \mathfrak{F}_\alpha\right] = E\left[\int_\alpha^\beta X_s^2 ds | \mathfrak{F}_\alpha\right] \qquad (7.31)$$

which implies the following very important equality:

$$E\left[\left(\int_\alpha^\beta X_s dB_s\right)^2\right] = E\left[\int_\alpha^\beta X_s^2 ds\right] \qquad (7.32)$$

having a profound geometrical meaning, as we will see in a moment. The linear map:

$$I : S^2(\alpha, \beta) \to L^2(\Omega, \mathfrak{F}_\beta, P), \quad X \to I(X) \stackrel{def}{=} \int_\alpha^\beta X_s dB_s \qquad (7.33)$$

satisfies the following:

$$||I(X)||_{L^2(\Omega, \mathfrak{F}_\beta, P)}^2 = E\left[\int_\alpha^\beta X_s^2 ds\right] \qquad (7.34)$$

The right hand side of this equality is a double integral in dt and $P(d\omega)$ of the function

$$(t, \omega) \to X_t^2(\omega) \qquad (7.35)$$

so that it has the form of a square norm in a L^2-like space which we define in the following:

Definition 7.4 We denote $M^2(\alpha, \beta)$ the set of equivalence classes of **progressively measurable** processes such that:

$$E\left[\int_\alpha^\beta X_s^2 ds\right] < +\infty \qquad (7.36)$$

where, as before, we say that two processes X and Y are equivalent if:

$$P\left(\int_\alpha^\beta |X_t - Y_t| dt = 0\right) = 1 \qquad (7.37)$$

$M^2(\alpha, \beta)$ is a Hilbert space, subspace of $L^2\left((\alpha, \beta) \times \Omega, \mathcal{B}(\alpha, \beta) \otimes \mathcal{F}_\beta, \lambda \otimes P\right)$, where λ is the Lebesgue measure.

We have thus built an **isometry**, called **Itô isometry**:

$$||I(X)||^2_{L^2(\Omega, \mathcal{F}_\beta, P)} = ||X||^2_{M^2(\alpha, \beta)} \qquad (7.38)$$

Itô isometry is not only a beautiful geometric identity, but it is the cornerstone of the extension of the Itô integral to more general processes.

7.3.2 First Extension of the Itô Integral

So far, we have defined the Itô integral for simple processes, through Eq. (7.20). In this section, we extend it to processes in $M^2(\alpha, \beta)$. We have learned the the map:

$$I : \mathcal{S}^2(\alpha, \beta) \subset M^2(\alpha, \beta) \to L^2(\Omega, \mathcal{F}_\beta, P), \quad X \to I(X) \overset{def}{=} \int_\alpha^\beta X_s dB_s \quad (7.39)$$

is linear and isometric:

$$||I(X)||^2_{L^2(\Omega, \mathcal{F}_\beta, P)} = ||X||^2_{M^2(\alpha, \beta)} \qquad (7.40)$$

and thus it is bounded. The intuitive idea is to extend the Itô integral to more general processes using approximating sequences of simple processes. This is possible since it can be shown [1] that $\mathcal{S}^2(\alpha, \beta)$ is a **dense** subset of $M^2(\alpha, \beta)$, that is, for each process $X \in M^2(\alpha, \beta)$, there exists a sequence $\left\{Y^{(n)}\right\}_n$ of simple processes in $\mathcal{S}^2(\alpha, \beta)$ such that:

$$X = \lim_{n \to +\infty} Y^{(n)} \qquad (7.41)$$

where the above limit is in L^2-sense, that is:

$$E\left[\int_\alpha^\beta |Y_s^{(n)} - X_s|^2 ds\right] \overset{n\to+\infty}{\to} 0 \qquad (7.42)$$

Thanks to the well-known Hahn-Banach bounded extension theorem [2, 3], the bounded linear functional (7.39) can be extended from square-integral simple processes to general square-integrable processes, leading to the following:

Definition 7.5 We call **stochastic integral** of a process $X \in M^2(\alpha, \beta)$, and we denote $\int_\alpha^\beta X_s dB_s$, the element of $L^2(\Omega, \mathcal{F}_\beta, P)$:

$$\int_\alpha^\beta X_s dB_s \overset{def}{=} \lim_{n\to\infty} \int_\alpha^\beta Y_s^{(n)} dB_s \qquad (7.43)$$

where $\left\{Y^{(n)}\right\}_n \subset \mathcal{S}^2(\alpha, \beta)$ is any sequence of simple square-integrable processes converging to X in $M^2(\alpha, \beta)$. The above written limit is meant in the topology of $L^2(\Omega, \mathcal{F}_\beta, P)$.

The map I becomes thus an isometry between Hilbert spaces:

$$I : M^2(\alpha, \beta) \to L^2(\Omega, \mathcal{F}_\beta, P), \quad X \to I(X) \overset{def}{=} \int_\alpha^\beta X_s dB_s \qquad (7.44)$$

The properties of the restriction to $\mathcal{S}^2(\alpha, \beta)$ guarantee that all the properties of $I(X)$ discussed above, including the ones about expectations and conditional expectations, hold for each $X \in M^2(\alpha, \beta)$.

7.3.3 Second Extension of the Itô Integral

In this section, we extend the Itô integral beyond $M^2(\alpha, \beta)$. We limit ourselves to outlining the extension procedure, which relies on suitable approximating sequences. Technical details can be found in [1].

Definition 7.6 We let $\Lambda^2(\alpha, \beta)$ be the set of equivalence classes of **progressively measurable** processes such that:

$$P\left(\int_\alpha^\beta |X_s|^2 ds < +\infty\right) = 1 \qquad (7.45)$$

where we identify two processes X and Y if:

$$P\left(\int_\alpha^\beta |X_t - Y_t| dt = 0\right) = 1 \qquad (7.46)$$

Naturally $M^2(\alpha, \beta) \subset \Lambda^2(\alpha, \beta)$.

It is possible to show [1] that for each process $X \in \Lambda^2(\alpha, \beta)$ there exists a sequence of simple processes $\{Y^{(n)}\}_n$ in $\Lambda^2(\alpha, \beta)$, such that:

$$\int_\alpha^\beta |Y_s^{(n)} - X_s|^2 ds \xrightarrow{n \to +\infty} 0 \qquad (7.47)$$

where the limit is meant in probability. It can then be proved [1] that the sequence:

$$\left\{ \int_\alpha^\beta Y_s^{(n)} d B_s \right\}_n \qquad (7.48)$$

converges in probability to a random variable which depends on X but not on the approximating sequence. Such random variable is the Itô integral of the process X:

$$\int_\alpha^\beta X_s d B_s \overset{def}{=} \lim_{n \to +\infty} \int_\alpha^\beta Y_s^{(n)} d B_s \qquad (7.49)$$

where again the convergence is in probability. It is quite simple to show that, if $X \in M^2(\alpha, \beta)$ this definition coincides to the one given above. When dealing with processes outside $M^2(\alpha, \beta)$, care has to be taken since properties involving expectations and conditional expectations are no longer valid.

So far, our discussion has been concerned with proving the existence of the Itô integral. Of course, practical applications require its evaluation, which will be explored in the reminder of the chapter. It can be shown [4] that, whenever a process X is continuous, its Itô integral is the limit in probability of the following sequence of Riemann sums:

$$\sum_{i=0}^{n-1} X_{t_i} (B_{t_{i+1}} - B_{t_i}) \qquad (7.50)$$

as the width $|t_{i+1} - t_i|$ of the partition tends to 0. An application of this formula is proposed in Problem 7.2.

7.3.4 The Itô Integral as a Function of Time

So far we have defined the Itô integral of a process X over an interval with fixed extrema α and β as a map producing a random variable $\int_\alpha^\beta X_s d B_s$. Allowing β to vary, such random variable is promoted to a stochastic process.

A central object of stochastic calculus is the following process:

$$I(t) \overset{def}{=} \int_0^t X_s d B_s \qquad (7.51)$$

where $X \in \Lambda^2(0, T)$, and the instant t is the interval $[0, T]$.

The following very simple but important property holds for $s < t$:

$$I(t) = I(s) + \int_s^t X_s d B_s \tag{7.52}$$

Moreover, $I(t)$ is $\mathcal{F}_t$-measurable for all t; we already know that this is the case if X is a simple process. In the general case, if $\{Y^{(n)}\}_n$ is a sequence of simple processes approximating in probability X in $\Lambda^2(0, T)$, and $I_n(t) = \{\int_0^t Y_s^{(n)} d B_s\}_n$, then $I_n(t)$ is $\mathcal{F}_t$-measurable. Since $I_n(t)$ converges to $I(t)$ in probability, it is possible to extract a subsequence converging almost surely to $I(t)$, which is thus $\mathcal{F}_t$-measurable.

It is possible to show [1] that each equivalence class $X \in \Lambda^2(0, T)$ contains a representative such that $I(t)$ is **continuous**.

In the particular case $X \in M^2(0, T)$, we already know that $I(t)$ is square-integrable and:

$$E[I(t)|\mathcal{F}_s] = I(s) + E\left[\int_s^t X_s d B_s | \mathcal{F}_s\right] = I(s) \tag{7.53}$$

so that $I(t)$ is a **square-integrable martingale**.

Despite its somewhat abstract appearance, the Itô integral can be evaluated exactly for a broad class of processes. One extremely important exact computation of a Itô integral is the Wiener integral, widely used in physics and finance, as well as in the theory of complex systems [5].

7.3.5   The Wiener Integral

A special case happens when the integrand is a deterministic function of time. Let $f :$ $[0, T] \to \mathbb{R}$ be a square-integrable real valued function $f \in L^2(0, T)$. The process:

$$(t, \omega) \to f(t) \tag{7.54}$$

independent of ω, is in $M^2(0, T)$. In such case:

$$I(t) = \int_0^t f(s) d B_s \tag{7.55}$$

is called **Wiener integral** of f on $[0, t]$. We know, from Eqs. (7.27) and (7.32) that:

$$E[I(t)] = 0, \quad E[I(t)^2] = \int_0^t f^2(s) ds \tag{7.56}$$

If f is piecewise constant:

$$f(s) = \sum_{i=0}^{n-1} c_i \, 1_{[t_i, t_{i+1})}(s) \tag{7.57}$$

then:

$$I(t) = \sum_{i=0; t_i < t, t_{i+1} < t}^{n-1} c_i \left(B_{t_{i+1}} - B_{t_i} \right) \tag{7.58}$$

and thus $I(t)$ is normal, being a linear combination of normal random variables. This holds also for $(I(t_1), \ldots, I(t_n))$, which turns out to be normal. This property continues to hold in the limit in $M^2(0, T)$, since the convergence in the sense of L^2 preserves the normal character of laws of random variables. We conclude that:

$$I(t) = \int_0^t f(s) dB_s \sim N\left(0, \int_0^t f^2(s) ds\right) \tag{7.59}$$

It is possible to exactly compute the covariance function of a Wiener integral, as discussed in Problem 7.6.

7.4   Stochastic Differential and Itô's Lemma

So far we have learned to define integrals of stochastic processes, both with respect to time and with respect to the Brownian motion. Before proceeding, let us pause for a second and briefly summarize the results we obtained. We have fixed the mathematical environment assigning a stochastic basis:

$$\left(\Omega, \mathcal{F}, \{\mathcal{F}_t\}_{t \geq 0}, P\right) \tag{7.60}$$

in the usual hypothesis, that is with a right-continuous filtration such that $\mathcal{F}_t$ contains all the elements of $\mathcal{F}$ whose probability is zero. Then we have started from a one-dimensional continuous Brownian motion $\{B_t\}_t$ with increments independent of the past. Given a time interval $[0, T]$,

1. for all $F \in \Lambda^1(0, T)$ we have built a process $\int_0^t F_s ds$, for $t \in [0, T]$, continuous, and thus progressively measurable. Moreover the trajectories of such process are integrable and also square-integrable being continuous on the compact interval $[0, T]$, so that:

$$\left\{\int_0^t F_s ds\right\}_{0 \leq t \leq T} \in \Lambda^1(0, T) \cap \Lambda^2(0, T) = \Lambda^2(0, T) \tag{7.61}$$

2. for all $G \in \Lambda^2(0, T)$ we have built a process $\int_0^t G_s dB_s$, for $t \in [0, T]$, continuous, and thus progressively measurable. Moreover the trajectories of such process are integrable and also square-integrable being continuous on the compact interval $[0, T]$, so that:

$$\left\{ \int_0^t G_s dB_s \right\}_{0 \le t \le T} \in \Lambda^1(0, T) \cap \Lambda^2(0, T) = \Lambda^2(0, T) \qquad (7.62)$$

Definition 7.7 Let $\{X_t\}_{t \ge 0}$ be a process such that, $\forall t \in [0, T]$, the following equality holds:

$$X_t = X_0 + \int_0^t F_s ds + \int_0^t G_s dB_s \qquad (7.63)$$

X_0 being a $\mathcal{F}_0$-measurable random variable, $F \in \Lambda^1(0, T)$ and $G \in \Lambda^2(0, T)$. Then, we say that $\{X_t\}_{t \ge 0}$ is an **Itô process** or, equivalently, that $\{X_t\}_{t \ge 0}$ has **stochastic differential**:

$$dX_t = F_t dt + G_t dB_t \qquad (7.64)$$

We stress that the stochastic differential is not a new mathematical object, but it provides a mere rewriting of a stochastic integral. However, the differential formalism is much easier to deal with, due to an extremely important result which is the cornerstone of stochastic calculus, the **Itô formula** or Itô's lemma. For a proof we refer to [6].

Theorem 7.1 (Itô's lemma) *Let $X^{(i)}$, $i = 1, \ldots, m$ be a collection of Itô processes with differentials:*

$$dX_t^{(i)} = F_t^{(i)} dt + G_t^{(i)} dB_t \qquad (7.65)$$

with $F^{(i)} \in \Lambda^1(0, T)$ and $G^{(i)} \in \Lambda^2(0, T)$. Let also $f : \mathbb{R}^m \times \mathbb{R}^+ \to \mathbb{R}$ be a measurable function, continuous at every point $(\mathbf{x}, t)$, $\mathbf{x} = (x_1, \ldots, x_m)$, twice continuously differentiable in $\mathbf{x}$, and once in t. Then, writing $X_t \overset{def}{=} (X_t^{(1)}, \ldots, X_t^{(m)})$, the process $Y_t \overset{def}{=} f(X_t, t)$ is an Itô process with differential:

$$dY_t = f_t(X_t, t)dt + \sum_{i=1}^m f_{x_i}(X_t, t)dX_t^{(i)} + \frac{1}{2} \sum_{i,j=1}^m f_{x_i x_j}(X_t, t) G_t^{(i)} G_t^{(j)} dt \qquad (7.66)$$

where the subscripts t and x_i denote derivatives with respect to t and x_i.

that is:

$$dY_t =$$
$$= \left(f_t(X_t, t) + \sum_{i=1}^m f_{x_i}(X_t, t) F_t^{(i)} + \frac{1}{2} \sum_{i,j=1}^m f_{x_i x_j}(X_t, t) G_t^{(i)} G_t^{(j)} \right) dt +$$
$$+ \left(\sum_{i=1}^m f_{x_i}(X_t, t) G_t^{(i)} \right) dB_t \qquad (7.67)$$

7.5 Extension to the Multidimensional Case

We conclude this section sketching the extension of the stochastic calculus formalism to the multidimensional case. The generalization is straightforward: nothing actually changes but a slight modification of the notations.

The starting point is, as usual, a stochastic basis in the usual hypothesis where a continuous d-dimensional Brownian motion with increments independent of the past is assigned.

The processes F_t of the previous paragraphs take now values in $\mathbb{R}^m$ while the processes G_t take values in $\mathbb{R}^{m \times d}$. We say that F_t belongs to $\Lambda_m^1(0, T)$, $T > 0$ if $F_{i,t}$ belongs to $\Lambda^1(0, T)$ for all $i = 1, \ldots, m$. In the same way. we say that G_t belongs to $\Lambda_{m,d}^2(0, T)$ (respectively $M_{m,d}^2(0, T)$) if $G_{ij,t}$ belongs to $\Lambda^2(0, T)$ (respectively $M^2(0, T)$) for all $i = 1, \ldots, m$, $j = 1, \ldots d$.

The time integral $\int_0^t F_s ds$ is defined as the vector of components $\int_0^t F_{i,s} ds$, while $\int_0^t G_s dB_s$ is defined as the vector of components $\sum_{j=1}^d \int_0^t G_{ij,s} dB_{j,s}$. Itô isometry becomes:

$$E\left[\left| \int_0^t G_s dB_s \right|^2 \right] = \int_0^t E\left[|G_s|^2 \right] ds \qquad (7.68)$$

and holds whenever $G_t \in M_{m,d}^2(0, T)$. We observe that $|\ |$ in the left hand side denotes the norm in $\mathbb{R}^m$, while in the right hand side it denotes the norm in $\mathbb{R}^{m \times d}$.

If we can write:

$$X_t = X_0 + \int_0^t F_s ds + \int_0^t G_s dB_s \qquad (7.69)$$

with X_0 $\mathcal{F}_0$-measurable we say that X is an Itô process, or, equivalently, that X has stochastic differential:

$$dX_t = F_t dt + G_t dB_t \qquad (7.70)$$

The Itô's formula can be generalized as follows (see [6]):

Theorem 7.2 (Multidimensional Itô formula) *Let X be a process taking values in $\mathbb{R}^m$ with stochastic differential:*

$$dX_t = F_t dt + G_t dB_t \qquad (7.71)$$

with $F_t \in \Lambda_m^1(0, T)$ and $G \in \Lambda_{m,d}^2(0, T)$. We let also $f : \mathbb{R}^m \times \mathbb{R}^+ \to \mathbb{R}$ be a measurable function, continuous in every point $(\mathbf{x}, t)$, $\mathbf{x} = (x_1, \ldots, x_m)$, continuously differentiable twice in $\mathbf{x}$ and once in t. Then the process $Y_t \overset{def}{=} f(X_t, t)$ admits stochastic differential:

$$dY_t = f_t(X_t, t)dt + \sum_{i=1}^m f_{x_i}(X_t, t)dX_{i,t} + \frac{1}{2}\sum_{i,j=1}^m f_{x_i x_j}(X_t, t)\sum_{h=1}^d G_{ih,t}G_{jh,t}dt \qquad (7.72)$$

Itô's lemma has a vast breadth of application and use in many branches of applied mathematics and physics and underlies the formalism of stochastic differential equations, forming the subject of the next chapter. Stochastic differential equations are powerful generalizations of ordinary differential equations, that include randomness in the modeling of physical processes.

Problems 7.3–7.6 propose applications of Itô's lemma, that can help the reader familiarizing with this powerful tool of stochastic calculus.

Before tackling the theory of SDEs, we conclude the chapter presenting the historically important Langevin equation.

7.6 The Langevin Equation

We consider again the motion of a pollen grain inside a glass of water, already introduced in Chap. 6. We can use the formalism of the stochastic calculus to provide a more general description of this kind of random motion. We would like to write down a Newton equation of motion for the pollen grain which takes into account, at a phenomenological level, the interaction between the grain and the water molecules. The presence of the water gives rise both to a velocity-dependent drag force of the form $-\zeta \mathbf{v}(t)$, arising from the viscosity of the water, described through a friction coefficient $\zeta > 0$, and to a random force, say $\mathbf{f}(t)$, representing the collisions with water molecules surrounding the grain at a given instant t.

The simplest model of these collisions uses the famous **white noise**, usually heuristically introduced requiring $\mathbf{f}(t)$ to have Gaussian distribution, zero mean and no memory of the past:

$$E[f_i(t)] = 0, \quad Cov(f_i(t), f_j(s)) \propto \delta_{ij}\delta(t-s), \quad i, j = 1, \ldots, 3 \qquad (7.73)$$

Physically, this means that the correlations decay faster than any time scale important in the physical description of the motion.

At a rigorous level, since the covariance is a Dirac delta distribution, some care has to be taken in defining $\mathbf{f}(t)$. We are going to show now that the properties (7.73) could characterize the *time derivative*, if existing, of the Brownian motion. To this aim, let us define:

$$w_{t,h} = \frac{B_{t+h} - B_t}{h} \qquad (7.74)$$

for a finite increment h. By inspection we see that $w_{t,h} \sim N(0, \frac{1}{h})$. Moreover, for $s < t$:

$$Cov\left(w_{t,h}w_{s,h}\right) = \frac{1}{h^2}(s + h - min(t, s + h)) \qquad (7.75)$$

Letting $h \to 0$, the above expression tends to 0 whenever $s \neq t$ and to ∞ in the special case $s = t$, justifying $\delta(t-s)$. The problem is that the trajectories of the

Brownian motion are not differentiable: this circumstance prevents to rigorously define $\mathbf{f}(t)$. The formalism of stochastic calculus permits to circumvent difficulties related with the ill-definition of $\mathbf{f}(t)$ by rigorously introducing a white noise term:

$$dt\,\mathbf{f}(t) \to d\mathbf{B}_t \qquad (7.76)$$

Using the notations $\mathbf{x}_t$ and $\mathbf{v}_t$, which are two stochastic processes with state space $\mathbb{R}^3$, to indicate the position and the velocity of the pollen grain at the instant t, we write the Newton equation in the form of a **Langevin equation**:

$$\begin{cases} d\mathbf{x}_t = \mathbf{v}_t dt \\ d\mathbf{v}_t = -\zeta\mathbf{v}_t dt + \sigma d\mathbf{B}_t \end{cases} \qquad (7.77)$$

The above equation, from a formal point of view, is simply a stochastic differential for a process $X_t = (\mathbf{x}_t, \mathbf{v}_t)$ with state space $\mathbb{R}^6$, having assigned a three-dimensional Brownian motion:

$$dX_t = -\mathcal{A}X_t dt + \mathcal{S}d B_t \qquad (7.78)$$

where $\mathcal{A}$ is a constant 6×6-matrix and $\mathcal{S}$ a constant 6×3 matrix.

Turning to the integral form, we have:

$$X_t = X_0 - \int_0^t \mathcal{A}X_s ds + \int_0^t \mathcal{S}d B_s \qquad (7.79)$$

We observe that the above formula is not a solution, but an equation, since the unknown process X also appears in the right hand side, and still has to be determined. In order to build up an explicit solution, we use the Ansatz:

$$X_t = e^{-\mathcal{A}t}U_t \qquad (7.80)$$

and apply the Itô formula to the function $f(\mathbf{x}, t) = e^{-\mathcal{A}t}\mathbf{x}$ (observing that the second derivatives with respect to $x_i x_j$ vanish). We get:

$$dX_t = d\left(e^{-\mathcal{A}t}U_t\right) = -\mathcal{A}e^{-\mathcal{A}t}U_t dt + e^{-\mathcal{A}t}dU_t = -\mathcal{A}X_t dt + e^{-\mathcal{A}t}dU_t \quad (7.81)$$

and thus, from a comparison with the equation of motion for X:

$$dU_t = e^{\mathcal{A}t}\mathcal{S}d B_t \qquad (7.82)$$

introducing a deterministic initial condition:

$$X_0 = U_0 = (\mathbf{x}_0, \mathbf{v}_0) \qquad (7.83)$$

we have the explicit solution of the Langevin equation:

$$X_t = e^{-At}X_0 + \int_0^t e^{-A(t-s)}\mathcal{S}dB_s \tag{7.84}$$

describing the random motion of the pollen grain. In order to keep the notations simple, we turn now to the one-dimensional case, where:

$$A = \begin{pmatrix} 0 & -1 \\ 0 & \zeta \end{pmatrix}, \quad \mathcal{S} = \begin{pmatrix} 0 \\ \sigma \end{pmatrix} \tag{7.85}$$

It is simple to show, by a direct calculation, that:

$$A^n = \begin{pmatrix} 0 & -\zeta^{n-1} \\ 0 & \zeta^n \end{pmatrix} \tag{7.86}$$

whence the exponential e^{-At} is:

$$e^{-At} = \begin{pmatrix} 1 & (1-e^{-\zeta t})/\zeta \\ 0 & e^{-\zeta t} \end{pmatrix} \tag{7.87}$$

The solution takes the form:

$$\begin{pmatrix} x_t \\ v_t \end{pmatrix} = \begin{pmatrix} 1 & (1-e^{-\zeta t})/\zeta \\ 0 & e^{-\zeta t} \end{pmatrix} \begin{pmatrix} x_0 \\ v_0 \end{pmatrix} + \int_0^t \begin{pmatrix} 1 & (1-e^{-\zeta(t-s)})/\zeta \\ 0 & e^{-\zeta(t-s)} \end{pmatrix} \begin{pmatrix} 0 \\ \sigma \end{pmatrix} dB_s \tag{7.88}$$

that is:

$$\begin{cases} x_t = x_0 + \frac{(1-e^{-\zeta t})}{\zeta}v_0 + \int_0^t \frac{(1-e^{-\zeta(t-s)})}{\zeta}\sigma dB_s \\ v_t = e^{-\zeta t}v_0 + \int_0^t e^{-\zeta(t-s)}\sigma dB_s \end{cases} \tag{7.89}$$

Both position and velocity are a sum of a deterministic contribution, on a time scale $1/\zeta$ related to the viscous drag, and of a random contribution, arising from the collisions with water molecules. These random terms have the form of Wiener integrals of functions of the variable s, square-integrable on the interval $(0, t)$, for each value of t. We know from (7.59) that one-dimensional Wiener integrals are **normal** random variables with zero expectation and variance equal to the time integral of the square of the deterministic integrand. Thus:

$$E[x_t] = x_0 + \frac{(1-e^{-\zeta t})}{\zeta}v_0, \quad Var(x_t) = \frac{\sigma^2}{\zeta^2}\int_0^t \left(1-e^{-\zeta(t-s)}\right)^2 ds \tag{7.90}$$

Explicitly:

$$Var(x_t) = \frac{\sigma^2 t}{\zeta^2} + \frac{\sigma^2}{2\zeta^3}(-3 + 4e^{-\zeta t} - e^{-2\zeta t}) \tag{7.91}$$

and:

$$E[v_t] = e^{-\zeta t}v_0, \quad Var(v_t) = \sigma^2 \int_0^t \left(e^{-\zeta(t-s)}\right)^2 ds = \sigma^2 \frac{1 - e^{-2\zeta t}}{2\zeta} \tag{7.92}$$

Let us write explicitly the probability density for the velocity of the pollen grain at the instant t:

$$p(v,t) = \left(\frac{2\zeta}{2\pi\sigma^2(1 - e^{-2\zeta t})}\right)^{1/2} \exp\left(-\frac{2\zeta(v - e^{-\zeta t}v_0)^2}{2\sigma^2(1 - e^{-2\zeta t})}\right) \tag{7.93}$$

In the realm of liquid state theory, a very important object is the autocorrelation of velocity:

$$C_v(\tau) = E\left[v_{t+\tau}v_t\right] \tag{7.94}$$

For an explicit calculation of this dynamic correlation function we need the following important property of Wiener integrals:

$$\left\{I_t = \int_0^t f(s)dB_s\right\}_t, \quad f \in L^2(0,T), \quad 0 \le t \le T \tag{7.95}$$

namely:

$$E[I_t I_{t'}] = \int_0^{\min(t,t')} f^2(s)ds \tag{7.96}$$

The proof is left as an exercise (Problem 7.6).

Putting all together, using (7.96) for the Wiener integral in v_t, we have:

$$E[v_t v_{t'}] = v_0^2 e^{-\zeta(t+t')} + e^{-\zeta(t+t')}\sigma^2 \int_0^{\min(t,t')} e^{2\zeta s}ds \tag{7.97}$$

that is:

$$E[v_t v_{t'}] = v_0^2 e^{-\zeta(t+t')} + e^{-\zeta(t+t')}\frac{\sigma^2}{2\zeta}\left(e^{2\zeta\min(t,t')} - 1\right) \tag{7.98}$$

or, equivalently:

$$E[v_t v_{t'}] = \frac{\sigma^2}{2\zeta}e^{-\zeta|t-t'|} + \left(v_0^2 - \frac{\sigma^2}{2\zeta}\right)e^{-\zeta(t+t')} \tag{7.99}$$

so that, for $\tau > 0$:

$$C_v(\tau) = E\left[v_{t+\tau}v_t\right] = \frac{\sigma^2}{2\zeta}e^{-\zeta\tau} + \left(v_0^2 - \frac{\sigma^2}{2\zeta}\right)e^{-\zeta(2t+\tau)} \tag{7.100}$$

Since we have an analytic solution, we can investigate the limit $t \to +\infty$. From a physical point of view, this means that we study the random processes when $t \gg 1/\zeta$, the latter playing the role of a *relaxation* time. We have:

$$E[v_t] = e^{-\zeta t} v_0 \overset{t \to +\infty}{\longrightarrow} 0, \quad Var(v_t) = \sigma^2 \frac{1 - e^{-2\zeta t}}{2\zeta} \overset{t \to +\infty}{\longrightarrow} \frac{\sigma^2}{2\zeta} \tag{7.101}$$

so that, as can be checked by considering the characteristic function, v_t converges in law to a random variable $N(0, \frac{\sigma^2}{2\zeta})$, with density:

$$p_\infty(v) = \left(\frac{2\zeta}{2\pi\sigma^2} \right)^{1/2} \exp\left(-\frac{2\zeta v^2}{2\sigma^2} \right) \tag{7.102}$$

If the glass of water is kept at a temperature T, the above mentioned convergence in law corresponds to a thermalization of the pollen grain, suggesting to postulate the relation:

$$\frac{2\zeta}{\sigma^2} = \frac{m}{k_B T} \tag{7.103}$$

m being the mass of the grain. In this way, the *equilibrium density* has the typical Maxwell-Boltzmann form (in one dimension):

$$p_\infty(v) = \left(\frac{m}{2\pi k_B T} \right)^{1/2} \exp\left(-\frac{mv^2}{2k_B T} \right) \tag{7.104}$$

Moreover, for $t \gg 1/\zeta$, the autocorrelation of the velocity has the exponential form:

$$C_v(\tau) = E\left[v_{t+\tau} v_t \right] \overset{t \gg 1/\zeta}{\simeq} \frac{k_B T}{m} e^{-\zeta\tau}, \quad \tau \geq 0 \tag{7.105}$$

As far as the position is concerned, we have, in one dimension, the important result:

$$E[x_t] = x_0 + \frac{(1 - e^{-\zeta t})}{\zeta} v_0 \overset{t \to +\infty}{\longrightarrow} x_0 + \zeta^{-1} v_0 \tag{7.106}$$

which supports the interpretation of ζ^{-1} as *relaxation time*. The variance provides information about the quadratic mean displacement:

$$Var(x_t) \overset{t \to +\infty}{\simeq} \frac{\sigma^2}{\zeta^2} t = 2\frac{k_B T}{\zeta m} t \tag{7.107}$$

growing linearly with time.

7.6.1 Ohm's Law

The Langevin equation can be easily generalized to study the motion of a particle subject to a constant external force, that can be of gravitational or electrostatic origin, leading to a suggestive derivation of Ohm's law. The Langevin equation for the velocity becomes:

$$\begin{cases} dv_t = \frac{F}{m} dt - \zeta v_t \, dt + \sigma \, dB_t \\ v_0 = v_0 \quad a.s. \end{cases} \tag{7.108}$$

To solve this equation, we introduce an auxiliary process:

$$w_t = f(v_t, t) = e^{\zeta t} v_t + \frac{F}{m\zeta} \left(1 - e^{\zeta t}\right) \tag{7.109}$$

Applying the Itô formula, we find that:

$$dw_t = \sigma e^{\zeta t} \, dB_t \ \rightarrow \ w_t = v_0 + \int_0^t ds \, \sigma \, e^{\zeta s} \, dB_s \tag{7.110}$$

Inverting Eq. (7.109) we obtain:

$$v_t = v_0 e^{-\zeta t} + \frac{F}{m\zeta}(1 - e^{-\zeta t}) + \int_0^t ds \, \sigma \, e^{\zeta(s-t)} \, dB_s \tag{7.111}$$

whence:

$$v_t \sim N\left(v_0 e^{-\zeta t} + \frac{F}{m\zeta}\left(1 - e^{-\zeta t}\right), \frac{\sigma^2}{2\zeta}\left(1 - e^{-2\zeta t}\right)\right) \tag{7.112}$$

In particular:

$$E[v_t] = v_0 e^{-\zeta t} + \frac{F}{m\zeta}\left(1 - e^{-\zeta t}\right) \overset{t \to +\infty}{\longrightarrow} \frac{F}{m\zeta} \tag{7.113}$$

The most noticeable aspect of Eq. (7.113) is that the asymptotic *drift* velocity is proportional to the external force.

In particular, for a collection of N non-interacting particles obeying this Langevin equation with an electrostatic force $F = qE$ one finds:

$$E[v_t] \to \frac{q}{m\gamma} E \quad J = \frac{1}{V} \sum_{i=1}^N E[v_t] = \frac{q^2 n}{m\gamma} E \tag{7.114}$$

where $n = N/V$ is the electron density.

7.7 Further Readings

Readers interested in deepening their knowledge about the mathematical foundations of the stochastic integration can see, for example [1, 5–7]. Readers more interested in applications can refer, just to quote a few examples, to [5, 7]. Applications within statistical mechanics can be found, for example, in [8]. Finally, stochastic calculus in the realm of statistical field theory are presented in [9, 10].

Problems

7.1 Time integral of the Brownian motion
Given a one-dimensional Brownian motion continuous with increments independent from the past B_t, define:

$$X_t = \int_0^t B_s ds \tag{7.115}$$

Find $E[X_t]$ and $Var(X_t)$. What can we say about the average $\frac{X_T}{T}$.

7.2 Itô integral of the Brownian motion
Given a one-dimensional Brownian motion continuous with increments independent from the past B_t, compute:

$$\int_0^t B_s dB_s \tag{7.116}$$

starting from the definition of the Itô integral.

7.3 Integration by parts
Given a one-dimensional Brownian motion continuous with increments independent from the past B_t, consider two Itô processes X_t, Y_t with stochastic differentials:

$$X_t = f_1(X_t)dt + g_1(X_t)dB_t \quad Y_t = f_2(Y_t)dt + g_2(Y_t)dB_t$$

Show that:

$$d(X_t Y_t) = X_t dY_t + Y_t dX_t + g_1(X_t)g_2(Y_t)dt$$

7.4 Stochastic differentials
Given a one-dimensional Brownian motion continuous with increments independent from the past B_t, compute the stochastic differentials of the processes B_t^2 and $\sin(t + B_t)$.

7.5 Stochastic differential equations
Show that the equation:

$$dX_t = f(t)X_t dt + g(t)dB_t \quad X_0 = x_0$$

has the solution:

$$X_t = x_0 F(t) + F(t) \int_0^t F(-s)g(s)dB_s, \quad F(t) = \exp\left(\int_0^t f(s)ds\right)$$

7.6 Correlation functions of Wiener processes
Show that:

$$E[I_t I_{t'}] = \int_0^{\min(t,t')} f^2(s)ds \qquad (7.117)$$

holds for Wiener integrals:

$$\left\{ I_t = \int_0^t f(s)dB_s \right\}_t, \quad f \in L^2(0,T), \quad 0 \le t \le T \qquad (7.118)$$

References

1. Baldi, P.: Equazioni Differenziali Stocastiche e Applicazioni. Pitagora Editori (2000)
2. Rudin, W.: Real and Complex Analysis. McGraw-Hill (1987)
3. Reed, M., Simon, B.: Functional Analysis. Academic Press (1980)
4. Kloeden, P.E., Platen, E.: Numerical Solution of Stochastic Differential Equations. Springer (1992)
5. Capasso, V., Bakstein, D.: An Introduction to Continuous Time Stochastic Processes. Birkhäuser (2005)
6. Karatzas, I., Shreve, S.E.: Brownian Motion and Stochastic Calculus. Springer (2005)
7. Oksendal, B.: Stochastic Differential Equation. Springer (2000)
8. Schwabl, F., Brewer, W.D.: Statistical Mechanics. Springer (2006)
9. Parisi, G.: Statistical Field Theory. Avalon Publishing (1988)
10. Zinn-Justin, J.: Quantum Field Theory and Critical Phenomena. Oxford Science Publication (2002)

Chapter 8
Stochastic Differential Equations

Abstract In this chapter we introduce the formalism of stochastic differential equations (SDE). After an introduction stressing their importance as generalizations of ordinary differential equations (ODE), we discuss existence and uniqueness of their solutions and we prove the Markov property. This leads us to a deep connection with the theory of partial differential equations (PDE), which will emerge naturally when computing time derivatives of averages of the processes. In particular we will introduce the generalized heat equation, as well as the more general Feynman-Kac equation, underlying the path integral formalism and the Schrödinger equation in imaginary time. Moreover, studying the time evolution of the transition probability of the processes will lead us to the Kolmogorov equations. A special case is provided by the Liouville equation, the cornerstone of classical statistical mechanics.

Keywords Stochastic differential equations · Chapman-Kolmogorov equation
Fokker-Planck equation · Geometric brownian motion · Brownian bridge
Feynman-Kac equation · Kakutani representation

8.1 General Introduction to Stochastic Differential Equations

Let us briefly summarize where do we stand: in the previous chapter we have learnt to define integrals of given processes with respect to time and with respect to the brownian motion and we have introduced the notation of stochastic differential:

$$dX_t = F_t dt + G_t dB_t \tag{8.1}$$

in which the coefficients F_t and G_t are given stochastic processes. On the other hand, while studying the generalization of Newton equation in presence of random forces, we have brought into stage a differential in which the coefficients were functions of the unknown solution itself. This drives us into the formalism of stochastic differential equations. In this chapter we define on a rigorous mathematical basis a class of stochastic differential equations generalizing the Langevin equation and we present, without proof, the theorems about existence and unicity of solutions. The proofs involve sophisticated mathematical instruments, the presentation of which is beyond the aim of this book.

Let's fix a time interval $[0, T]$: any instant of time that will appear from now on lies in this interval. Let moreover b and σ be measurable functions:

$$b : \mathbb{R}^m \times [0, T] \to \mathbb{R}^m, \quad \sigma : \mathbb{R}^m \times [0, T] \to \mathbb{R}^{m \times d} \tag{8.2}$$

that we call **drift** and **diffusion coefficient** respectively. We use b and σ to write a formal stochastic differential equation for an unknown process X_t taking values in $\mathbb{R}^m$:

$$\begin{cases} dX_t = b(X_t, t)dt + \sigma(X_t, t)dB_t \\ \quad X_u = \eta, \quad u \le t \end{cases} \tag{8.3}$$

η being a m-dimensional random variable.

The Eq. (8.3) is a powerful generalization of the Cauchy problem in the realm of ordinary differential equations (ODE), for which $\sigma \equiv 0$ and X_u non random. As the reader may remember from courses about basic calculus, the approach to the Cauchy problem starts from the definition of its solution. Then, properties of the function b ensuring existence and uniqueness of the solutions are identified and, finally, specific techniques to solve the problem are investigated in a well defined mathematical framework. We will follow the same path in the realm of SDE. We also comment that, as in the Cauchy problem, the differential dX_t is given in terms of functions evaluated at the precise instant t, i.e. no memory effects are included in the time evolution equation. Introducing memory effects, although very important, lies beyond the purpose of this book. Interested readers are deferred, for example, to the book [1], where a thorough discussion of the Langevin equation with memory is presented.

At this point we have written only a formal equality: we have not fixed a stochastic basis yet. Let us therefore rigorously define what do we mean when talking about a solution of (8.3).

Definition 8.1 We say that a process:

$$X = \left(\Omega, \mathcal{F}, \{\mathcal{F}_t\}_{t \in [0,T]}, \{X_t\}_{t \in [0,T]}, P \right) \tag{8.4}$$

is a solution of the **stochastic differential equation** (8.3) if:

1. $(\Omega, \mathcal{F}, \{\mathcal{F}_t\}_t, \{B_t\}_t, P)$ is a continuous d-dimensional brownian motion with increments independent of the past defined inside a stochastic basis satisfying the usual hypotheses;
2. η is $\mathcal{F}_u$-measurable;
3. for all $t \in [u, T]$, we have:

$$X_t = \eta + \int_u^t b(X_s, s)ds + \int_u^t \sigma(X_s, s)dB_s \tag{8.5}$$

We stress that it is implicitly assumed that the two integrals in the above equation are well defined.

The existence and the main properties of solutions of (8.3) will be discussed under the following working hypotheses, closely resembling the ones required in the realm of ordinary differential equations:

Definition 8.2 (*hypotheses (A)*) We say that b and σ satisfy the hypotheses (A) if they are measurable in $(\mathbf{x}, \mathbf{t})$ and if there exist $L > 0$ and $M > 0$ such that, for each $\mathbf{x}, \mathbf{y} \in \mathbb{R}^d$, and $t \in [0, T]]$, the following sublinear growth and global Lipschitz conditions holds:

$$|b(\mathbf{x}, t)| \leq M(1 + |\mathbf{x}|), \quad |\sigma(\mathbf{x}, t)| \leq M(1 + |\mathbf{x}|) \tag{8.6}$$

$$|b(\mathbf{x}, t) - b(\mathbf{y}, t)| \leq L|\mathbf{x} - \mathbf{y}|, \quad |\sigma(\mathbf{x}, t) - \sigma(\mathbf{y}, t)| \leq L|\mathbf{x} - \mathbf{y}| \tag{8.7}$$

Remark 8.1 The above hypotheses (A) are quite natural: in the special case $\sigma = 0$ we expect to recover the formalism of ordinary differential equations.

Under such hypotheses the following global existence and uniqueness theorem can be proved [2–5]:

Theorem 8.1 *Given a stochastic basis in the usual hypotheses (defined in Sect. 6.5 in Chap. 6), where a continuous d-dimensional brownian motion with increments independent of the past is defined, if η is a m-dimensional random variable $\mathcal{F}_u$-measurable square-integrable, $E[|\eta|^2] < +\infty$ and if the hypotheses (A) hold, there exists a process $X \in M^2(u, T)$ such that:*

$$X_t = \eta + \int_u^t b(X_s, s)ds + \int_u^t \sigma(X_s, s)dB_s \tag{8.8}$$

Moreover, if another process X' satisfies Eq. (8.8), then:

$$P\left(X_t = X'_t, \ \forall t \in [u, T]\right) = 1 \tag{8.9}$$

Along with existence and uniqueness, in this context it is possible to study continuous dependence on initial data, by focussing on the situation when (8.3) has a deterministic initial condition $\eta = \mathbf{x}$ almost surely. We denote $X_t^{\mathbf{x},s}$ the solution of:

$$\begin{cases} dX_t = b(X_t, t)dt + \sigma(X_t, t)dB_t \\ \quad\quad X_s = \mathbf{x}, \quad \mathbf{x} \in \mathbb{R}^m \end{cases} \tag{8.10}$$

We state without proof this result [4], concerning continuous dependence on initial data:

Theorem 8.2 *Under hypotheses (A) there exists a collection of m-dimensional random variables* $\{Z_{\mathbf{x},s}(t)\}_{\mathbf{x},s,t}$, *with* $\mathbf{x} \in \mathbb{R}^m$, $0 \le s \le t \le T$ *such that.*

1. *the map* $(\mathbf{x}, s, t) \to Z_{\mathbf{x},s}(t)$ *is continuous for each* ω;
2. $Z_{\mathbf{x},s}(t) = X_t^{\mathbf{x},s}$ *almost surely for all* $(\mathbf{x}, s, t)$.

The family of processes $(t, \omega) \to X_t^{\mathbf{x},s}(\omega)$, thus, almost surely depends continuously on the initial position $\mathbf{x}$, on the initial instant s, and on the time variable t.

8.2 Stochastic Differential Equations and Markov Processes

The importance of the family of processes $X_t^{\mathbf{x},s}$, introduced at the end of the last paragraph, is manifold. On one hand, they allow to build up the solution also in the general case of random starting point. Indeed, it is possible to prove [4] that the process

$$(t, \omega) \to X_t(\omega) = X_t^{\eta(\omega),s}(\omega), \quad t \ge s \tag{8.11}$$

is a solution of:

$$\begin{cases} dX_t = b(X_t, t)dt + \sigma(X_t, t)dB_t \\ \quad X_s = \eta, \quad s \le t \end{cases} \tag{8.12}$$

if η is $\mathcal{F}_u$-measurable and square integrable.

Remark 8.2 The notation $X_t^{\eta(\omega),s}(\omega)$, widely used in the literature about SDE, represents simply a composition of measurable functions:

$$\omega \to \eta(\omega) \to X_t^{\eta(\omega),s}(\omega) \tag{8.13}$$

As the processes $X_t^{\mathbf{x},s}$ allow to easily build the solution for a generic random variable η, they will be the focus of our attention from now on.

In particular, a corollary of the result (8.11) is the following composition property:

$$X_t^{\mathbf{x},s}(\omega) = X_t^{X_u^{\mathbf{x},s}(\omega),u}(\omega), \quad s \le u \le t \tag{8.14}$$

whose meaning is clarified in Fig. 8.1: $X_t^{\mathbf{x},s}(\omega)$ starts from $\mathbf{x}$ at time s and visits the random point $X_u^{\mathbf{x},s}(\omega)$ at intermediate time u.

Starting from (8.14), we will show that the processes $X_t^{\mathbf{x},s}$ are Markov processes, satisfying the celebrated Chapman-Kolmogorov equation:

$$p(A, t \mid \mathbf{x}, s) = \int_{\mathbb{R}^d} p(A, t \mid \mathbf{y}, u) \, p(d\mathbf{y}, u \mid \mathbf{x}, s) \tag{8.15}$$

where:

$$p(A, t \mid \mathbf{x}, s) = P\left(X_t^{\mathbf{x},s} \in A\right) \tag{8.16}$$

denotes the probability that the process $X_t^{\mathbf{x},s}$, which started in $\mathbf{x}$ at time s, lies inside the set A at time t.

8.2.1 The Chapman-Kolmogorov Equation

Let $X_t^{\mathbf{x},s}$ be the solution of:

$$\begin{cases} dX_t = b(X_t, t)dt + \sigma(X_t, t)dB_t \\ \quad X_s = \mathbf{x}, \quad \mathbf{x} \in \mathbb{R}^m \end{cases} \tag{8.17}$$

If A is a Borel subset of $\mathbb{R}^m$ we can define the real valued function:

$$p(A, t \mid \mathbf{x}, s) \overset{def}{=} P(X_t^{\mathbf{x},s} \in A) = E\left[1_A(X_t^{\mathbf{x},s})\right] \tag{8.18}$$

The dependence on $\mathbf{x}$ is measurable as a consequence of the continuity in $(\mathbf{x}, s, t)$, and the dependence on A provides a probability measure, the law of the random variable $\omega \to X_t^{\mathbf{x},s}(\omega)$.

We are going to show now that p satisfies the Markov property for the process $X_t^{\mathbf{x},s}$:

$$P\left(X_t^{\mathbf{x},s} \in A \mid \mathcal{F}_u\right) = p(A, t \mid X_u^{\mathbf{x},s}, u) \tag{8.19}$$

To this aim, we need the relation (8.14), which we remind here:

$$X_t^{\mathbf{x},s}(\omega) = X_t^{X_u^{\mathbf{x},s}(\omega),u}(\omega), \quad s \le u \le t \tag{8.20}$$

Let's define the $\mathbf{x}$-dependent random variable:

$$\psi(\mathbf{x}, \omega) = 1_A(X_t^{\mathbf{x},u}(\omega)) \tag{8.21}$$

We observe that:

$$P\left(X_t^{\mathbf{x},s} \in A \mid \mathcal{F}_u\right) = E\left[1_A\left(X_t^{\mathbf{x},s}\right) \mid \mathcal{F}_u\right] = \tag{8.22}$$
$$= E\left[1_A\left(X_t^{X_u^{\mathbf{x},s},u}\right) \mid \mathcal{F}_u\right] = E\left[\psi(X_u^{\mathbf{x},s}, \cdot) \mid \mathcal{F}_u\right]$$

Now, the random variable:

$$\omega \to X_u^{\mathbf{x},s}(\omega) \tag{8.23}$$

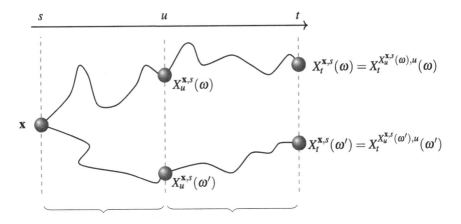

$X_u^{\mathbf{x},s}$ is $\mathcal{F}_u$-measurable $X_t^{\mathbf{x},u}$ is $\mathcal{F}_u$-independent

Fig. 8.1 Pictorial representation of the rationale behind Eq. (8.14). The process $(t, \omega) \to X_t^{\mathbf{x},s}(\omega)$ starts, at time s, from the point $\mathbf{x}$. At a given intermediate time u, any position acts as the starting point for the future evolution

is $\mathcal{F}_u$-measurable, while the random variable:

$$\omega \to \psi(\mathbf{x}, \omega) = 1_A(X_t^{\mathbf{x},u}(\omega)) \qquad (8.24)$$

is independent of $\mathcal{F}_u$, since, intuitively, whatever happens before u doesn't matter having fixed the process in $\mathbf{x}$ at the time u. We can thus use Theorem 3.4 in Chap. 3 about conditional expectations:

$$E\left[\psi(X_u^{\mathbf{x},s}, \cdot)|\mathcal{F}_u\right] = \Phi\left(X_u^{\mathbf{x},s}\right), \quad \Phi(\mathbf{y}) = E\left[\psi(\mathbf{y}, \cdot)\right] \qquad (8.25)$$

We have:

$$\Phi(\mathbf{y}) = E\left[\psi(\mathbf{y}, \cdot)\right] = E\left[1_A\left(X_t^{\mathbf{y},u}\right)\right] = p(A, t \mid \mathbf{y}, u) \qquad (8.26)$$

Putting all together, we have proved the Markov property:

$$P\left(X_t^{\mathbf{x},s} \in A|\mathcal{F}_u\right) = p(A, t \mid X_u^{\mathbf{x},s}, u) \qquad (8.27)$$

We need to show now that Chapman-Kolmogorov property holds, which is quite simple:

$$p(A, t \mid \mathbf{x}, s) = E\left[1_A(X_t^{\mathbf{x},s})\right] = E\left[E\left[1_A(X_t^{\mathbf{x},s})|\mathcal{F}_u\right]\right] = \qquad (8.28)$$
$$= E\left[p(A, t \mid X_u^{\mathbf{x},s}, u)\right] = \int_{\mathbb{R}^d} p(A, t \mid \mathbf{y}, u) \, p(d\mathbf{y}, u \mid \mathbf{x}, s)$$

We have thus learned that $\{X_t^{\mathbf{x},s}\}$ is a Markov process with initial instant s, initial law $\delta_{\mathbf{x}}$ and transition function p. We already know that such process is continuous; moreover, since $X_t^{\mathbf{x},s}$ is continuous in $(\mathbf{x}, s, t)$, the map:

$$(\mathbf{x}, t) \rightarrow \int_{\mathbb{R}^d} d\mathbf{y}\, f(\mathbf{y})\, p(d\mathbf{y}, t + h \,|\, \mathbf{x}, t) = E\left[X_{t+h}^{\mathbf{x},t}\right] \tag{8.29}$$

is continuous for any function f continuous and bounded. In the language of Markov processes theory, this is called **Feller property**, which, together with the continuity of the process, makes $\{X_t^{\mathbf{x},s}\}$ a **strong Markov process**.

8.3 Kolmogorov Equations

As we learnt in Sect. 6.9, every Markov process can be associated to an infinitesimal generator. Building over the result that the solutions of SDE are indeed Markov processes, we will now construct the infinitesimal generator for solutions of SDEs. This will deepen our exploration of the connection between two different branches of Mathematics: the theory of stochastic processes and the theory of partial differential equations. As anticipated in Sect. 6.9, we will learn how to build a Markov process starting from a partial differential equation. This will provide an extremely useful tool for numeric solutions of PDEs.

The central result is the correspondence between a stochastic differential equation:

$$dX_t = b(X_t, t)dt + \sigma(X_t, t)dB_t \tag{8.30}$$

and a differential operator:

$$\mathcal{L}_t = \frac{1}{2} \sum_{i,j=1}^{d} a_{i,j}(\mathbf{x}, t) \frac{\partial^2}{\partial x_i \partial x_j} + \sum_i b_i(\mathbf{x}, t) \frac{\partial}{\partial x_i} \tag{8.31}$$

where the matrix $a(\mathbf{x}, t) = \sigma(\mathbf{x}, t)\, (\sigma(\mathbf{x}, t))^T$ is positive semidefinite. Under some conditions, in fact, the solution of a differential equation of the form:

$$\frac{\partial}{\partial t}\phi(\mathbf{x}, t) = (\mathcal{L}_t \phi)(\mathbf{x}, t), \quad \phi(\mathbf{x}, 0) = f(\mathbf{x}) \tag{8.32}$$

can be expressed as:

$$\phi(\mathbf{x}, t) = E[f(X_t)] \tag{8.33}$$

where:

$$dX_t = b(X_t, t)dt + \sigma(X_t, t)dB_t, \quad X_0 = \mathbf{x} \tag{8.34}$$

Remark 8.3 The special case $f(\mathbf{x}) = \delta(\mathbf{x} - \mathbf{x}_0)$ leads to the so-called **fundamental solution** of the PDE. Under the natural hypothesis that transition function is absolutely continuous:

$$p(d\mathbf{y}, t \mid \mathbf{x}, s) = d\mathbf{y}\, q(\mathbf{y}, t \mid \mathbf{x}, s) \tag{8.35}$$

we have:

$$E\left[\delta\left(X_t^{\mathbf{x},s} - \mathbf{x}_0\right)\right] = q(\mathbf{x}_0, t \mid \mathbf{x}, s) \tag{8.36}$$

meaning that the transition probability density q, for the *backward* transition:

$$(\mathbf{x}, s) \rightarrow (\mathbf{x}_0, t) \tag{8.37}$$

We use the term *backward* since $\mathbf{x}_0$ is the *starting point* for the PDE.

For example, consider the case of the m-dimensionsl brownian motion: a simple calculation shows that, if $X_t = \mathbf{x} + B_t$, then the function

$$\phi(\mathbf{x}, t) = E[f(X_t)] = \int d\mathbf{y}\, f(\mathbf{y}) \frac{1}{(2\pi t)^{m/2}} \exp\left(-\frac{|\mathbf{y} - \mathbf{x}|^2}{2t}\right) \tag{8.38}$$

is the solution of the heat equation:

$$\frac{\partial}{\partial t}\phi(\mathbf{x}, t) = \left(\frac{1}{2}\nabla^2 \phi\right)(\mathbf{x}, t), \quad \phi(\mathbf{x}, 0) = f(\mathbf{x}) \tag{8.39}$$

This result is usually expressed saying that the differential operator $\mathcal{L}_t = \frac{1}{2}\nabla^2$ is the **infinitesimal generator of the brownian motion**.

Now, more generally, we consider a measurable real-valued function f limited and smooth, more presicely $C^2(\mathbb{R}^m)$ with bounded derivatives, and define the map:

$$(T_{s,t}f)(\mathbf{x}) \stackrel{def}{=} E\left[f\left(X_t^{\mathbf{x},s}\right)\right] = \int_{\mathbb{R}^d} f(\mathbf{y})\, p(d\mathbf{y}, t \mid \mathbf{x}, s) \tag{8.40}$$

where, as in the previous paragraph, $\{X_t^{\mathbf{x},s}\}$ satisfies the equation:

$$\begin{cases} dX_t = b(X_t, t)dt + \sigma(X_t, t)dB_t \\ \quad X_s = \mathbf{x}, \quad \mathbf{x} \in \mathbb{R}^m \end{cases} \tag{8.41}$$

We assume moreover that hypotheses (A) hold.

We apply now Itô formula to the process $f\left(X_t^{\mathbf{x},s}\right)$, obtaining:

$$df\left(X_t^{\mathbf{x},s}\right) = \sum_{i=1}^{m} \frac{\partial f\left(X_t^{\mathbf{x},s}\right)}{\partial x_i}\left(dX_t^{\mathbf{x},s}\right)_i + \frac{1}{2}\sum_{i,j=1}^{m}\frac{\partial^2 f\left(X_t^{\mathbf{x},s}\right)}{\partial x_i \partial x_j} a_{i,j}(X_t^{\mathbf{x},s},t)dt \quad (8.42)$$

where:

$$a = \sigma\,\sigma^T \qquad\qquad (8.43)$$

We define now:

$$(\mathcal{L}_t f)(\mathbf{x}) = \frac{1}{2}\sum_{i,j=1}^{m} a_{i,j}(\mathbf{x},t)\frac{\partial^2 f}{\partial x_i \partial x_j}(\mathbf{x}) + \sum_{i=1}^{m} b_i(\mathbf{x},t)\frac{\partial f}{\partial x_i}(\mathbf{x}) \qquad (8.44)$$

so that:

$$df\left(X_t^{\mathbf{x},s}\right) = (\mathcal{L}_t f)(X_t^{\mathbf{x},s})dt + \sum_{i=1}^{m}\sum_{j=1}^{d}\frac{\partial f\left(X_t^{\mathbf{x},s}\right)}{\partial x_i}\sigma_{i,j}\left(X_t^{\mathbf{x},s},t\right)dB_{j,t} \quad (8.45)$$

By construction, the derivatives of f are limited and σ has a sublinear growth; moreover we know that $X_t^{\mathbf{x},s}$ belongs to $M^2(s,T)$, which implies that the coefficient of the differential of the brownian motion belongs to $M^2(s,T)$. We can thus be sure that the Itô integral has zero mean. We can thus write:

$$E\left[f\left(X_t^{\mathbf{x},s}\right)\right] = f(\mathbf{x}) + \int_s^t du\, E\left[(\mathcal{L}_u f)(X_u^{\mathbf{x},s})\right] \qquad (8.46)$$

that is:

$$\left(T_{s,t}f\right)(\mathbf{x}) = f(\mathbf{x}) + \int_s^t du\left(T_{s,u}\circ\mathcal{L}_u f\right)(\mathbf{x}) \qquad (8.47)$$

We observe that, in (8.47), if $T_{s,t}$ commutes with $\mathcal{L}_u$, we are lead to a differential equation for the function:

$$(\mathbf{x},t,s) \rightarrow \left(T_{s,t}f\right)(\mathbf{x}) \qquad (8.48)$$

which is precisely the central result of this paragraph. The investigation of the conditions making $T_{s,t}$ commute with $\mathcal{L}_u$, is beyond the scope of this book.

Interestingly, it is found that differentiation with respect to the initial time s allows to weaken the hypothesis. The following important result in fact can be shown [3, 4]:

Theorem 8.3 (Backward Kolmogorov equation) *If the hypotheses (A) hold and $\forall R > 0$, there exists $\lambda_R > 0$ such that:*

$$(a(\mathbf{x}, t)\,\mathbf{z}) \cdot \mathbf{z} \geq \lambda_R |\mathbf{z}|^2 \tag{8.49}$$

for all $(\mathbf{x}, t)$, $|\mathbf{x}| < R$, $0 \leq t \leq T$ and $\mathbf{z} \in \mathbb{R}^m$, then, defining $u^t(\mathbf{x}, s) \overset{def}{=} (T_{s,t} f)(\mathbf{x})$ for f limited and continuous, $u^t(\mathbf{x}, s)$ is the unique solution with polynomial growth on $[0, t)$ of the **Backward Kolmogorov equation**:

$$\begin{cases} \frac{\partial u}{\partial s} = -\mathcal{L}_s u \\ \lim_{s \to t^-} u(\mathbf{x}, t) = f(\mathbf{x}) \end{cases} \tag{8.50}$$

8.3.1 The Fokker-Planck Equation

Another very interesting point is to write down the equation of motion for the transition probability of the stochastic process $X_t^{\mathbf{x},s}$. For example, in the particular case of the brownian motion, the transition probability density:

$$p(\mathbf{y}, t \mid \mathbf{x}, s) = \frac{1}{(2\pi(t-s))^{m/2}} \exp\left(-\frac{|\mathbf{y} - \mathbf{x}|^2}{2(t-s)}\right) \tag{8.51}$$

satisfies the heat equation:

$$\frac{\partial}{\partial t} p(\mathbf{y}, t \mid \mathbf{x}, s) = \frac{1}{2}\nabla^2 p(\mathbf{y}, t \mid \mathbf{x}, s) \tag{8.52}$$

where the laplacian operator is meant with respect to the variable $\mathbf{y}$.

In general, let's assume that there exists a time dependent transition probability density:

$$p(A, t \mid \mathbf{x}, s) = \int_A d\mathbf{y}\, q(\mathbf{y}, t \mid \mathbf{x}, s), \quad t > s \tag{8.53}$$

We start from the basic expression:

$$E\left[f\left(X_t^{\mathbf{x},s}\right)\right] = f(\mathbf{x}) + \int_s^t du\, E\left[(\mathcal{L}_u f)(X_u^{\mathbf{x},s})\right] \tag{8.54}$$

where, as before, the function f is assumed to be measurable and limited, and $C^2(\mathbb{R}^d)$ with limited derivatives. Explicitly we have:

$$\int_{\mathbb{R}^d} d\mathbf{y}\, f(\mathbf{y}) q(\mathbf{y}, t \mid \mathbf{x}, s) = f(\mathbf{x}) + \tag{8.55}$$

$$\int_s^t du \int_{\mathbb{R}^d} d\mathbf{y}\, q(\mathbf{y}, u \mid \mathbf{x}, s) \left(\tfrac{1}{2} \sum_{i,j=1}^d a_{i,j}(\mathbf{y}, u) \partial_{i,j}^2 f(\mathbf{y}) + \sum_i b_i(\mathbf{y}, u) \partial_i f(\mathbf{y}) \right)$$

If the transition probability density is differentiable with respect to t for $t > s$ and if we can integrate by parts, we get:

$$\int_{\mathbb{R}^d} d\mathbf{y}\, f(\mathbf{y}) \left(\frac{\partial}{\partial t} - \frac{1}{2} \sum_{i,j=1}^d \partial_{i,j}^2 a_{i,j}(\mathbf{y}, t) + \sum_i \partial_i b_i(\mathbf{y}, t) \right) q(\mathbf{y}, t \mid \mathbf{x}, s) = 0$$

$$\tag{8.56}$$

Since such equation holds for any f under regularity assumptions discussed in [3], we are driven to the celebrated **Fokker-Planck equation** or **forward Kolmogorov equation**, providing the equation of motion of the transition probability density:

$$\frac{\partial}{\partial t} q(\mathbf{y}, t \mid \mathbf{x}, s) =$$

$$= \frac{1}{2} \sum_{i,j} \frac{\partial^2}{\partial y_i \partial y_j} \left(a_{i,j}(\mathbf{y}, t) q(\mathbf{y}, t \mid \mathbf{x}, s) \right) - \sum_i \frac{\partial}{\partial y_i} \left(b_i(\mathbf{y}, t) q(\mathbf{y}, t \mid \mathbf{x}, s) \right)$$

$$\tag{8.57}$$

Remark 8.4 In the deterministic case $a \equiv 0$, the Fokker-Planck equation reduces to the celebrated **Liouville equation**

$$\frac{\partial}{\partial t} q = - \sum_i \frac{\partial}{\partial y_i} (b_i\, q) \tag{8.58}$$

which is a cornerstone of classical statistical mechanics, in the special case when b is the vector field defined by the Hamiltonian of a physical system. In Problem 8.5 we will expand the formalism.

8.4 Important Examples

Before completing our exploration of the theory of SDEs, and in particular on their connection to PDEs, let us now pause for a moment, and take the opportunity to present some examples taken from applied science.

8.4.1 Geometric Brownian Motion

Let's consider the following equation in one dimension:

$$\begin{cases} dX_t = bX_t dt + \sigma X_t dB_t \\ \quad X_0 = x, \quad t \geq 0 \end{cases} \tag{8.59}$$

where b, σ and x are non-negative constants.

We will show now that the solution is the famous *geometric brownian motion*:

$$X_t = x \cdot \exp\left(\left(b - \frac{\sigma^2}{2}\right)t + \sigma B_t\right) \tag{8.60}$$

The parameter x is the initial value of the quantity X_t which always remains positive. In general such process is used to model the temporal evolution of prices in financial markets. In the case $\sigma = 0$, the evolution is *risk-less*, the constant b playing the role of *rate of increase*. The constant σ, usually called *volatility*, introduce *risk* in the temporal evolution, the term σB_t governing fluctuations in a price typical of a financial market.

Let's show that the above process actually satisfies the differential equation (8.73). To do this, we write $X_t = f(t, B_t)$ where:

$$f(t, y) = x \cdot \exp\left(\left(b - \frac{\sigma^2}{2}\right)t + \sigma y\right) \tag{8.61}$$

and apply Itô formula, obtaining:

$$\begin{aligned} dX_t &= \frac{\partial f}{\partial t} dt + \frac{\partial f}{\partial y} dB_t + \frac{1}{2} \frac{\partial^2 f}{\partial y^2} dt = \\ &= \left(b - \frac{\sigma^2}{2}\right) X_t dt + \sigma X_t dB_t + \frac{1}{2}\sigma^2 X_t dt = \\ &= bX_t dt + \sigma X_t dB_t \end{aligned} \tag{8.62}$$

which is what we wanted to show.

8.4.2 Brownian Bridge

Let's now consider the following equation:

$$\begin{cases} dX_t = -\frac{X_t}{1-t} dt + dB_t \\ \quad X_0 = 0, \quad 0 \leq t < 1 \end{cases} \tag{8.63}$$

We will show that the solution is:

$$X_t = (1 - t) \int_0^t \frac{dB_s}{1 - s} \tag{8.64}$$

Before doing this, let's discuss some general properties of this process. The first observation is that:

$$X_0 = X_1 = 0 \tag{8.65}$$

Moreover, the process is the product of a function depending only on time and a Wiener integral. This means that:

$$E[X_t] = 0 \tag{8.66}$$

and:

$$Var(X_t) = (1 - t)^2 \int_0^t ds \frac{1}{(1 - s)^2} = t(1 - t) \tag{8.67}$$

so that:

$$X_t \sim N(0, t(1 - t)) \tag{8.68}$$

This process is called *brownian bridge*.

We are going now to verify that the brownian bridge X_t actually satisfies the above written stochastic differential equations. For this purpose, we write:

$$X_t = f(t, Y_t), \quad f(t, x) = (1 - t)x, \quad Y_t = \int_0^t \frac{dB_s}{1 - s} \tag{8.69}$$

Itô formula implies that:

$$dX_t = -Y_t dt + (1 - t)dY_t = -\frac{X_t}{1 - t}dt + (1 - t)\frac{dB_t}{1 - t} \tag{8.70}$$

which is precisely what we wanted to show.

It is straightforward to generalize the brownian bridge to a process starting at a and arriving in b. The stochastic differential equation is the following:

$$\begin{cases} dX_t = \frac{b - X_t}{1 - t}dt + dB_t \\ X_0 = a, \quad 0 \le t < 1 \end{cases} \tag{8.71}$$

and the solution is:

$$X_t = (1 - t)a + tb + (1 - t) \int_0^t \frac{dB_s}{1 - s} \tag{8.72}$$

8.4.3 Langevin Equation in a Force Field

In this section we will discuss the Langevin equation at the presence of a force field [6, 7]. Apart from its intrinsic interest, this equation in indeed very important: on one hand, it offers a model of approach to equilibrium in a very natural way. On the other hand, it is the foundation of Smart Monte Carlo sampling, which is a powerful generalization of the Metropolis technique that we have introduced in Chaps. 4 and 5.

Let us consider the following equation, again in one dimension for simplicity:

$$\begin{cases} dX_t = \phi(X_t)dt + \sigma \, dB_t \\ \quad X_0 = x, \quad 0 \le t \end{cases} \tag{8.73}$$

where $\phi : \mathbb{R} \to \mathbb{R}$ is a function satisfying the conditions given in the previous chapter about existence and uniqueness of solutions of stochastic differential equations. On the other hand, $\sigma > 0$ is a constant.

Analytical solutions of this equations are, in general, unknown. Nevertheless, we can learn something about the process by studying the related Fokker-Planck equation, assuming that the transition probability density exists:

$$\frac{\partial}{\partial t}q(y,t\,|\,x,0) = \sigma^2 \frac{1}{2}\frac{\partial^2}{\partial y^2}q(y,t\,|\,x,0) - \frac{\partial}{\partial y}\left(\phi(y)q(y,t\,|\,x,0)\right) \tag{8.74}$$

Let's first look for a *time-independent* solution of the form:

$$q(y,t\,|\,x,0) \propto \exp\left(-\Phi(y)\right) \tag{8.75}$$

A solution of this form actually exists provided that:

$$\phi(y) = -\frac{\sigma^2}{2}\frac{\partial}{\partial y}\Phi(y) \tag{8.76}$$

Let's thus define:

$$q_0(y) = \frac{\exp\left(-\Phi(y)\right)}{\int_{\mathbb{R}} dy' \, \exp\left(-\Phi(y')\right)} \tag{8.77}$$

and look for time-dependent solutions of the Fokker-Planck equation, which we write in the form:

$$q(y,t\,|\,x,0) = \sqrt{q_0(y)}\,\tilde{\psi}(y,t) \equiv \psi_0(y)\,\tilde{\psi}(y,t) \tag{8.78}$$

If we substitute the above Ansatz in the Fokker-Planck equation we obtain:

$$-\partial_t\tilde{\psi}(y,t) = -\frac{1}{2}\sigma^2\partial_{yy}^2\tilde{\psi}(y,t) + \frac{1}{2}\sigma^2\left(\frac{\partial_{yy}^2\psi_0(y)}{\psi_0(y)}\right)\tilde{\psi}(y,t) \tag{8.79}$$

We define:

$$\frac{1}{2}\sigma^2 = \lambda = \frac{\hbar^2}{2m} \tag{8.80}$$

and the above equation takes the form of an imaginary time Schrödinger equation related to a Fokker-Planck hamiltonian with a local potential, of the form:

$$\hat{\mathcal{H}}_{FP} = -\lambda\partial^2_{yy} + \left(\frac{\lambda\partial^2_{yy}\psi_0(y)}{\psi_0(y)}\right) = -\lambda\partial^2_{yy} + \left(\frac{1}{4}(\partial_y\Phi(y))^2 - \frac{1}{2}\partial^2_y\Phi(y)\right) \tag{8.81}$$

It is immediate to observe that:

$$\hat{\mathcal{H}}_{FP}\psi_0 = 0 \tag{8.82}$$

that is ψ_0 is an eigenfunction of $\hat{\mathcal{H}}_{FP}$ relative to the eigenvalue 0. Moreover, ψ_0 is strictly positive, which means that it has to be the ground state of the Fokker-Planck hamiltonian, with zero energy [8].

The Schrödinger-like equation:

$$-\partial_t\tilde{\psi}(y,t) = \left(\hat{\mathcal{H}}_{FP}\tilde{\psi}\right)(y,t) \tag{8.83}$$

has general solution:

$$\tilde{\psi}(y,t) = \left(\exp\left(-t\hat{\mathcal{H}}_{FP}\right)f\right)(y,t) \tag{8.84}$$

f being any initial condition.

Since the ground state has zero energy, the excited states energies have to be positive, guaranteeing that, for any choice of the initial condition f, not orthogonal to ψ_0:

$$\tilde{\psi}(y,t) = \left(\exp\left(-t\hat{\mathcal{H}}_{FP}\right)f\right)(y,t) \xrightarrow{t\to+\infty} \psi_0(y) \tag{8.85}$$

a part from an unessential multiplicative constant.

Putting all together, we see that any solution of the Fokker-Planck equation related to the Langevin equation in a force field converges, in the limit $t \to +\infty$ to the *equilibrium probability density*:

$$q_0(y) = \frac{\exp\left(-\Phi(y)\right)}{\int_{\mathbb{R}} dy' \exp\left(-\Phi(y')\right)} \tag{8.86}$$

This is a very interesting model in which the phenomenon of approach to equilibrium appears: the stochastic motion described by the equation

$$dX_t = \phi(X_t)dt + \sigma dB_t \tag{8.87}$$

approaches a stationary asymptotic probability density independent of the initial condition: a Boltzmann weight related to the *potential energy* $\Phi(y)$ where:

$$\phi(y) = -\frac{\sigma^2}{2}\frac{\partial}{\partial y}\Phi(y) \tag{8.88}$$

This result is very useful also in the realm of numerical simulations, yielding the foundations of the so-called **smart Monte Carlo** or **Langevin Monte Carlo** method. Suppose we wish to sample the probability density:

$$p_0(y) = \frac{\exp(-\Phi(y))}{\int_{\mathbb{R}} dy' \exp(-\Phi(y'))} \tag{8.89}$$

If we were able to generate realizations of the solution of the stochastic differential equation:

$$\begin{cases} dX_t = \phi(X_t)dt + \sigma dB_t \\ \quad X_0 = x, \quad 0 \le t \end{cases} \tag{8.90}$$

then, after an equilibration transient, our simulation would yield the desired sampling. Although the solution cannot be found exactly, we can introduce an integration time step Δt and simulate the approximated solution:

$$X_{t+\Delta t} = X_t + \phi(X_t)\Delta t + \sigma(B_{t+\Delta t} - B_t) \tag{8.91}$$

Given $X_t = x^\star$, the random variable $X_{t+\Delta t}$ follows a law $N(x^\star + \phi(x^\star)\Delta t, \sigma^2 \Delta t)$. This is also an example of Euler-Maruyama method for the numerical solution of SDEs [9].

8.5   Feynman-Kac Equation

We are going now to explore further the connection between stochastic differential equations and partial differential equations. We have learned in previous chapters to relate the equation:

$$dX_t = b(X_t, t)dt + \sigma(X_t, t)dB_t \tag{8.92}$$

to the differential operator:

$$\mathcal{L}_t = \frac{1}{2} \sum_{ij} a_{ij}(x,t) \frac{\partial^2}{\partial x_i \partial x_j} + \sum_i b_i(x,t) \frac{\partial}{\partial x_i}$$

where $a = \sigma \sigma^T$. In this section we will deal with a more general class of PDEs of the form:

$$\begin{cases} -\partial_t u(\mathbf{x},t) + \mathcal{L}_t u(\mathbf{x},t) - V(\mathbf{x})u(\mathbf{x},t) = f(\mathbf{x},t) & (\mathbf{x},t) \in \mathbb{R}^n \times (0,T) \\ u(\mathbf{x},0) = \phi(\mathbf{x}) & \mathbf{x} \in \mathbb{R}^n \end{cases}$$

$$(8.93)$$

A reader familiar with Quantum Mechanics immediately realize the importance of such a generalization. The function $V(\mathbf{x})$ will have the interpretation of a time independent external potential in the particular case of the Schrödinger equation in imaginary time, $V(x)$ playing the role of the potential energy:

$$-\partial_t \Psi(x,t) = -\frac{1}{2} \Delta \Psi(x,t) + V(x)\Psi(x,t) \qquad (8.94)$$

Moreover, the term $f(\mathbf{x},t)$ appears, for example, in the heat equation with the role of a source term.

We will learn that also in this case the solution of the PDE can be represented as an average of a suitable functional of the process X_t. This connection will introduce naturally the path integral of the function $V(\mathbf{x})$ on the process X_t.

Let's fix some working hypotheses:

Definition 8.3 (*hypotheses B*) We will say that the operator $\mathcal{L}_t$:

$$\mathcal{L}_t = \frac{1}{2} \sum_{ij} a_{ij}(x,t) \frac{\partial^2}{\partial x_i \partial x_j} + \sum_i b_i(x,t) \frac{\partial}{\partial x_i}$$

satisfies hypothesis B if the functions $a_{ij}(x,t)$ and $b_i(x,t)$:

1. have sub-linear growth, that is there $\exists M$ such that:

$$|b_i(x,t)| \le M(1+|x|)$$
$$|a_{ij}(x,t)| \le M(1+|x|)$$

$\forall x \in \mathbb{R}^n, t \in (0,T)$.

2. satisfy Lipschitz condition, that is there $\exists L$ such that:

$$|b_i(x,t) - b_i(y,t)| \le L|x-y|$$
$$|a_{ij}(x,t) - a_{ij}(y,t)| \le L|x-y|$$

$\forall x, y \in \mathbb{R}^n, t \in (0,T)$.

In the realm of the theory of partial differential equations, the following existence and uniqueness theorem can be proved [10]:

Theorem 8.4 *Let $V : \mathbb{R}^n \to \mathbb{R}$, $f : \mathbb{R}^n \times (0, T) \to \mathbb{R}$ and $\phi : \mathbb{R}^n \to \mathbb{R}$ continuous functions and $\mathcal{L}_t$ the above mentioned differential operator statisfying hypothesis B and elliptic, that is there $\exists \Lambda$ such that:*

$$\sum_{ij} x_i\, a_{ij}(x, t)\, x_j \geq \Lambda |\mathbf{x}|^2 \quad \forall (\mathbf{x}, t) \in \mathbb{R}^n \times (0, T)$$

Then the parabolic partial differential equation:

$$\begin{cases} -\partial_t u(\mathbf{x}, t) + \mathcal{L}_t u(\mathbf{x}, t) - V(\mathbf{x}) u(\mathbf{x}, t) = f(\mathbf{x}, t) & (\mathbf{x}, t) \in \mathbb{R}^n \times (0, T) \\ u(\mathbf{x}, 0) = \phi(\mathbf{x}) \quad \mathbf{x} \in \mathbb{R}^n \end{cases} \tag{8.95}$$

has a unique solution $u(\mathbf{x}, t) \in C^2(\mathbb{R}^n \times (0, T))$

Theorem 8.5 (Feynman-Kac representation formula) *The solution $u(\mathbf{x}, t)$ of the parabolic partial differential equation:*

$$\begin{cases} -\partial_t u(\mathbf{x}, t) + \mathcal{L}_t u(\mathbf{x}, t) - V(\mathbf{x}) u(\mathbf{x}, t) = f(\mathbf{x}, t) & (\mathbf{x}, t) \in \mathbb{R}^n \times (0, T) \\ u(\mathbf{x}, 0) = \phi(\mathbf{x}) \quad \mathbf{x} \in \mathbb{R}^n \end{cases} \tag{8.96}$$

can be written as expectation of stochastic processes in the Feynman-Kac formula:

$$u(\mathbf{x}, T) = E[\phi(X_T)\, Z_T] - E\left[\int_0^T f(X_t, T - t)\, Z_t dt \right] \tag{8.97}$$

where X_t is the solution of the stochastic differential equation:

$$\begin{aligned} dX_t &= b(X_t, t)\, dt + \sigma(X_t, t)\, dB_t \\ X_0 &= \mathbf{x} \end{aligned} \tag{8.98}$$

where $a(\mathbf{x}, t) = \sigma(\mathbf{x}, t)\, \sigma(\mathbf{x}, t)^T$, and:

$$Z_t = e^{-\int_0^t V(X_s)ds} \tag{8.99}$$

is the path integral *of the function $V(x)$.*

Proof Let's define the stochastic process:

$$t \to \Phi_t = Z_t\, u(X_t, T - t) \tag{8.100}$$

where $u(x, t)$ is the unique solution of (8.96).

The first observation is that:

$$\Phi_0 = u(x, T), \quad \Phi_T = Z_T u(X_T, 0) = Z_T \phi(X_T) \qquad (8.101)$$

We are going now to evaluate $d\Phi_t$ observing that $\Phi_t = F(Z_t, X_t, t)$ where:

$$F(z, x, t) = z\, u(x, T - t) \qquad (8.102)$$

and using Itô formula. We observe that:

$$dZ_t = -Z_t\, V(X_t)\, dt \qquad (8.103)$$

as can be immediately verified from the very definition of Z_t. We have thus:

$$
\begin{aligned}
d\Phi_t = {} & -Z_t \partial_t u(X_t, T - t)dt + u(X_t, T - t)dZ_t + \\
& + Z_t \sum_i b_i(X_t, T - t)\partial_{x_i} u(X_t, T - t)dt + \\
& + Z_t \sum_{ij} \partial_{x_i} u(X_t, T - t)\sigma_{ij}(X_t, T - t)dB_{jt} + \qquad (8.104) \\
& + \frac{1}{2} Z_t \sum_{ij} a_{ij}(X_t, t)\partial_{x_i x_j} u(X_t, T - t)
\end{aligned}
$$

that is:

$$
\begin{aligned}
d\Phi_t = {} & Z_t \left(-\partial_t + \mathcal{L}_t\right) u(X_t, T - t)dt + u(X_t, T - t)dZ_t + \\
& + \sum_{ij} \partial_{x_i} u(X_t, T - t)\sigma_{ij}(X_t, T - t)dB_{jt} \qquad (8.105)
\end{aligned}
$$

We stress that the terms with $\partial_{zz} F$ and $\partial_{xz} F$ vanishes because the function F is linear in z and the Itô differential dZ_t does not contain dB_t.

Since, by construction, u is a solution of the partial differential equation (8.96), we get:

$$
\begin{aligned}
d\Phi_t = {} & Z_t \left(V(X_t)u(X_t, T - t) + f(X_t, T - t)\right)dt + u(X_t, T - t)dZ_t + \\
& \sum_{ij} \partial_{x_i} u(X_t, T - t)\sigma_{ij}(X_t, T - t)dB_{jt} = \qquad (8.106) \\
= {} & Z_t f(X_t, T - t)dt + \sum_{ij} \partial_{x_i} u(X_t, T - t)\sigma_{ij}(X_t, T - t)dB_{jt}
\end{aligned}
$$

where we have used the explicit expression for dZ_t.

We have thus:

$$\Phi_T - \Phi_0 = \int_0^T Z_t f(X_t, T - t)dt + \int_0^T \sum_{ij} \partial_{x_i} u(X_t, T - t)\sigma_{ij}(X_t, T - t)dB_{jt}$$

(8.107)

so that, taking the expectation of both members:

$$E[Z_T\phi(X_T)] - E[u(x, T)] = E\left[\int_0^T Z_t f(X_t, T - t)dt\right]$$ (8.108)

which gives the Feynmann-Kack representation:

$$u(x, T) = E[Z_T\phi(X_T)] - E\left[\int_0^T Z_t f(X_t, T - t)dt\right]$$ (8.109)

Let's specialize the Feynmann-Kac representation in the case of imaginary time Schrödinger equation:

$$-\partial_t \Psi(x, t) = -\frac{1}{2}\Delta\Psi(x, t) + V(x)\Psi(x, t)$$ (8.110)

$$\Psi(x, 0) = \phi(x)$$

We have:

$$u(x, T) = E\left[e^{-\int_0^T V(x+B_t)dt}\phi(x + B_T)\right]$$ (8.111)

where we have observed that, in such simple case $X_t = x + B_t$.

8.6   Kakutani Representation

We conclude our adventure with a very important application, which makes evident that the stochastic formalism is extremely powerful and can reach beyond the domain of partial differential equations describing some form of diffusion (Fig. 8.2).

Let's focus on the Poisson problem with Dirichlet boundary conditions:

$$\begin{cases} \frac{1}{2}\Delta u(x) = f(x) & x \in \Omega \\ u(x) = \phi(x) & x \in \partial\Omega \end{cases}$$ (8.112)

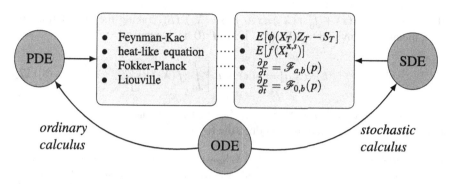

Fig. 8.2 Pictorial representation of the connection among Ordinary Differential Equations (ODE), Partial Differential Equations (PDE) and Stochastic Differential Equations (SDE)

where $\Omega \subset \mathbb{R}^n$ is a bounded open set. In one-dimension, the equation is:

$$\begin{cases} \frac{1}{2}u''(x) = f(x) & x \in (a, b) \\ u(a) = \phi_a \\ u(b) = \phi_b \end{cases} \tag{8.113}$$

and the solution has the simple form:

$$u(x) = \phi_a + \frac{x - a}{b - a}(\phi_b - \phi_a) + G(x) - \frac{x - a}{b - a}G(b) \tag{8.114}$$

where $G(x) = 2\int_a^x dx' \int_a^{x'} dx'' f(x'')$. On the other hand, when $n > 1$, the problem is much more difficult, and an analytical solution can be found only in a few special cases.

We already know that the differential operator $\mathcal{L} = \frac{1}{2}\Delta$ is related to the stochastic differential equation:

$$\begin{cases} dX_t = dB_t \\ X_0 = x \end{cases} \tag{8.115}$$

with solution $X_t = B_t + x$. For $t > 0$ the process X_t takes values *outside* the set Ω with probability $P(X_t \notin \Omega) \neq 0$, so that the process $Y_t = u(X_t)$ is well defined only if $f(x)$ and $u(x)$ can be extended to functions of class $C^2(\mathbb{R}^n)$. From now on we will assume that this is the case. Itô formula provides the following equality:

$$\begin{cases} dY_t = \nabla u(X_t)\,dB_t + \frac{1}{2}\Delta u(X_t, t)dt \\ Y_0 = u(x) \end{cases} \tag{8.116}$$

that is $u(X_t) = u(x) + \int_0^t f(X_s)\,ds + \int_0^t \nabla u(X_s)\,dB_s$. Taking the expectations of both members, provided that $\nabla u(X_s)$ is in $M^2(0, t)$, we get:

$$E[u(X_t)] = u(x) + E\left[\int_0^t f(X_s)\,ds\right] \tag{8.117}$$

We know the value of $E[u(X_t)]$ only if X_t lies on the boundary $\partial\Omega$, thanks to Dirichlet boundary conditions. We thus introduce the following:

Definition 8.4 (*first-pass instant*) If $\{X_t\}_{t\geq 0}$ is a stochastic process taking values inside a measurable space $(E, \mathcal{E})$, and $A \in \mathcal{E}$ is a measurable subset, the random variable:

$$\tau_A(\omega) = \inf\{t : X_t(\omega) \in A\} \tag{8.118}$$

is called *first-pass instant* of the process X_t in the set A.

and specialize (8.117) obtaining:

$$u(x) = E[\phi(X_{\tau_{\partial\Omega}})] - E\left[\int_0^{\tau_{\partial\Omega}} f(X_s)\,ds\right] \tag{8.119}$$

The interpretation of this result is simple: we can express the solution of the Poisson-Dirichlet problem as expectation of a suitable function of the process X_t.

The procedure we have followed is somehow euristic, since, in general, we cannot guarantee that $f(x)$ and $u(x)$ can be extended to functions of class $C^2(\mathbb{R}^n)$ and, moreover, we have introduced a substitution $t \to \tau_{\partial\Omega}$ in a non-rigorous way. Finally, we are not sure that $E[\tau_{\partial\Omega}] < \infty$.

A fully rigorous treatment, which goes beyond the aim of this book, relies on the the following two basic theorems, which we state without proof [10, 11]:

Theorem 8.6 (existence and uniqueness) *Let $\Omega \subset \mathbb{R}^n$ be a bounded open set, $c : \Omega \to [0, \infty)$ and $\phi : \partial\Omega \to \mathbb{R}$ functions satisfying lipschitz property and $\mathcal{L}$ a "time independent" differential operator of the form:*

$$\mathcal{L} = \frac{1}{2}\sum_{ij} a_{ij}(x)\frac{\partial^2}{\partial x_i \partial x_j} + \sum_i b_i(x)\frac{\partial}{\partial x_i}$$

satisfying hypothesis (B) and elliptic. The equation:

$$\begin{cases} \mathcal{L}u(x) - c(x)u(x) = f(x) & x \in \Omega \\ u(x) = \phi(x) & x \in \partial\Omega \end{cases} \tag{8.120}$$

has a unique solution $u(x) \in C^2(\Omega)$.

Theorem 8.7 (Kakutani formula) *Under the hypothesis of the previous theorem, the solution of the elliptic partial differential equation:*

$$\begin{cases} \mathcal{L}u(x) - c(x)u(x) = f(x) & x \in \Omega \\ u(x) = \phi(x) & x \in \partial\Omega \end{cases} \tag{8.121}$$

can be represented in the Kakutani form:

$$u(x) = E[Z_{\tau_{\partial\Omega}}\phi(X_{\tau_{\partial\Omega}})] - E\left[\int_0^{\tau_{\partial\Omega}} f(X_s)\, Z_s\, ds\right] \tag{8.122}$$

where X_t is the solution of the stochastic differential equation:

$$\begin{cases} dX_t = b(X_t)dt + \sigma(X_t)dB_t \\ X_0 = x \end{cases} \tag{8.123}$$

with $a(x) = \sigma(x)\sigma^T(x)$, $\tau_{\partial\Omega}$ is the first-pass instant of the process X_t in the boundary $\partial\Omega$ of the set Ω and:

$$Z_s = e^{-\int_0^s c(X_r)dr} \tag{8.124}$$

is the path integral of the function c.

8.7 Further Readings

The literature about stochastic differential equations and their applications is very vast. We defer the readers to the books cited in the main text of this chapter and references therein.

Problems

8.1 Population growth is a stochastic environment

Consider the following the equation for the size X_t of a population growing in a crowded and stochastic environment:

$$\begin{aligned} dX_t &= r\, X_t(K - X_t)dt + \beta X_t dB_t \\ X_0 &= x > 0 \end{aligned} \tag{8.125}$$

where $r \in \mathbb{R}$ represents the *quality of the environment*, $K > 0$ represents the *carrying capacity* of the environment and $\beta \in \mathbb{R}$ is a measure of the noise.

Show that the solution is:

$$X_t = \frac{\exp\left(\left(rK - \frac{\beta^2}{2}\right)t + \beta B_t\right)}{x^{-1} + r\int_0^t \exp\left(\left(rK - \frac{\beta^2}{2}\right)s + \beta B_s\right)dt} \tag{8.126}$$

8.2 RLC circuit with noise

Consider the following formal equation for a RLC circuit, without a generator, but subject to noise.

$$L\frac{d^2}{dt^2}Q(t) + R\frac{d}{dt}Q(t) + \frac{1}{C}Q(t) = \alpha f(t) \tag{8.127}$$

where $f(t)$ is a white noise term. Let's introduce a two-dimensional process $X_t = (X_{1,t}, X_{2,t})$ where $X_{1,t} = Q(t)$ and $X_{2,t} = \frac{d}{dt}Q(t)$. The problem becomes:

$$dX_{1,t} = X_{2,t}dt$$
$$dX_{2,t} = -\frac{R}{L}X_{2,t}dt - \frac{1}{LC}X_{1,t}dt + \frac{\alpha}{L}dB_t \tag{8.128}$$

or, in matrix notation:

$$dX_t = AX_tdt + \sigma dB_t, \quad A = \begin{pmatrix} 0 & 1 \\ -\frac{1}{LC} & -\frac{R}{L} \end{pmatrix}, \quad \sigma = \begin{pmatrix} 0 \\ \frac{\alpha}{L} \end{pmatrix} \tag{8.129}$$

Find the solution of this equation.

8.3 A function of the brownian motion

Find a function $f : \mathbb{R} \to \mathbb{R}^2$ such that $Y_t = f(B_t)$ satisfies the equation:

$$dY_{1,t} = -\frac{1}{2}Y_{1,t}dt - Y_{2,t}dB_t$$
$$dY_{2,t} = -\frac{1}{2}Y_{2,t}dt + Y_{1,t}dB_t \tag{8.130}$$

8.4 A calculation of the infinitesimal generator

Find the infinitesimal generator for the brownian bridge, starting from the equation:

$$dY_t = \frac{b - Y_t}{1 - t}dt + dB_t, \quad Y_0 = a \tag{8.131}$$

8.5 Liouville equation

Starting from classic $6N$-dimensional Hamilton equation:

$$\frac{d}{dt}\mathbf{x}(t) = \frac{\partial \mathcal{H}}{\partial \mathbf{p}}(\mathbf{x}(t), \mathbf{p}(t))$$
$$\frac{d}{dt}\mathbf{p}(t) = -\frac{\partial \mathcal{H}}{\partial \mathbf{x}}(\mathbf{x}(t), \mathbf{p}(t)) \tag{8.132}$$

write the Liouville equation for the probability density $\rho(\mathbf{x}, \mathbf{p}, t)$.

8.6 Cox-Ingersoll-Ross (CIR) model

In mathematical finance, the Cox-Ingersoll-Ross model describes the evolution of interest rates: the instantaneous interest rate X_t is assumed to satisfy the following stochastic differential equation:

$$dX_t = a(b - X_t)dt + \sigma\sqrt{X_t}dB_t, \quad X_0 = x_0 \tag{8.133}$$

Show that, if we choose $a = 2\beta$, $b = \frac{1}{8\beta}$, and $\sigma = 1$, it is possible to write $X_t = Y_t^2$, where Y_t satisfies the Langevin equation:

$$dY_t = -\beta Y_t dt + \frac{1}{2}dB_t \tag{8.134}$$

Write down the explicit solution in such case.

8.7 Wright-Fisher (WF) model

Consider the equation:

$$dX_t = (a + bX_t)dt + \sqrt{X_t(1 - X_t)}\,dB_t \quad X_0 = x_0$$

with $a, b, x_0 \in [0, 1]$. Show that, in the special case $a = \frac{1}{4}$ and $b = -\frac{1}{2}$, the solution is:

$$X_t = \sin^2\left(\frac{B_t}{2} + \arcsin\sqrt{x_0}\right) \tag{8.135}$$

References

1. Hansen, J.-P., McDonald, I.R.: Theory of Simple Liquids. Academic Press (2006)
2. Karatzas, I., Shreve, S.E.: Brownian Motion and Stochastic Calculus. Springer (2005)
3. Capasso, V., Bakstein, D.: An Introduction to Continuous Time Stochastic Processes. Birkhäuser (2005)
4. Baldi, P.: Equazioni Differenziali Stocastiche e Applicazioni. Pitagora Editori (2000)
5. Oksendal, B.: Stochastic Differential Equation. Springer (2000)
6. Zinn-Justin, J.: Quantum Field Theory and Critical Phenomena. Oxford Science Publication (2002)
7. Parisi, G.: Statistical Field Theory. Avalon Publishing (1988)
8. Feynman, R.P.: Statistical Mechanics: A Set of Lectures. Addison-Wesley (1972)
9. Kloeden, P.E., Platen, E.: Numerical Solution of Stochastic Differential Equations. Springer (1992)
10. Evans, L.C.: Partial Differential Equations. American Mathematical Society (2010)
11. Kakutani, S.: Two-dimensional Brownian Motion and Harmonic Functions. Proc. Imp. Acad. **20**(10), 706–714

Solutions

Problems of Chap. 1

1.1 If X is uniform in $(0, 1)$, we have:

$$E[X] = \int_0^1 dx\, x = \frac{1}{2} \tag{1}$$

and:

$$Var(X) = \int_0^1 dx \left(x - \frac{1}{2}\right)^2 = \frac{1}{12} \tag{2}$$

If $X \sim N(0, 1)$, we have:

$$E[X] = \int_{-\infty}^{+\infty} dx\, x\, \frac{1}{\sqrt{2\pi}} \exp\left(-\frac{x^2}{2}\right) = 0 \tag{3}$$

and:

$$Var(X) = \int_{-\infty}^{+\infty} dx\, x^2\, \frac{1}{\sqrt{2\pi}} \exp\left(-\frac{x^2}{2}\right) = 1 \tag{4}$$

If $X \sim B(n, p)$, we have:

$$
\begin{aligned}
E[X] &= \sum_{x=0}^{n} x\, \frac{n!}{(n-x)!x!} p^x (1-p)^{n-x} = \\
&= np \sum_{x=1}^{n} \frac{(n-1)!}{(n-x)!(x-1)!} p^{x-1}(1-p)^{(n-1)-(x-1)} = \\
&= np\, (p + (1-p))^{n-1} = np
\end{aligned}
\tag{5}
$$

© Springer International Publishing AG, part of Springer Nature 2018
E. Vitali et al., *Theory and Simulation of Random Phenomena*, UNITEXT
for Physics, https://doi.org/10.1007/978-3-319-90515-0

In order to evaluate the variance, let's start with:

$$
\begin{aligned}
E[X^2] &= \sum_{x=0}^{n} x^2 \frac{n!}{(n-x)!x!} p^x (1-p)^{n-x} = \\
&np \sum_{x=1}^{n} x \frac{(n-1)!}{(n-x)!(x-1)!} p^{x-1} (1-p)^{(n-1)-(x-1)} = \\
&= np \sum_{x=1}^{n} (x-1) \frac{(n-1)!}{(n-x)!(x-1)!} p^{x-1} (1-p)^{(n-1)-(x-1)} + np = \\
&= np(n-1)p \sum_{x=2}^{n} \frac{(n-2)!}{(n-x)!(x-2)!} p^{x-2} (1-p)^{(n-2)-(x-2)} + np = \\
&= np(n-1)p + np = n^2 p^2 + np(1-p)
\end{aligned}
\tag{6}
$$

We can now evaluate:

$$
Var(X) = E[X^2] - (E[X])^2 = np(1-p) \tag{7}
$$

There is also another approach in the case of the binomial distribution: let's consider the case $n = 1$ (the **Bernoulli distribution**). If $Y \sim B(1, p)$ we have:

$$
E[Y] = 1 \times p + 0 \times (1-p) = p, \tag{8}
$$

and:

$$
Var(Y) = (1-p)^2 \times p + (0-p)^2 \times (1-p) = (1-p)(p - p^2 + p^2) = p(1-p) \tag{9}
$$

Now, if $X \sim B(n, p)$, we can write $X = Y_1 + Y_2 + \cdots + Y_n$, where $Y_i \sim B(1, p)$ and the Y are independent. We thus find:

$$
E[X] = \sum_{i=1}^{n} E[Y_i] = np \tag{10}
$$

$$
Var(X) = \sum_{i=1}^{n} Var[Y_i] = np(1-p) \tag{11}
$$

If X is Poisson with parameter λ we have:

$$
\begin{aligned}
E[X] &= \sum_{x=0}^{+\infty} x \exp(-\lambda) \frac{\lambda^x}{x!} = \\
&= \exp(-\lambda) \lambda \sum_{x=1}^{+\infty} \frac{\lambda^{x-1}}{(x-1)!} = \lambda
\end{aligned}
\tag{12}
$$

In order to evaluate the variance, let's start with:

$$
\begin{aligned}
E[X^2] &= \sum_{x=0}^{+\infty} x^2 \exp(-\lambda)\frac{\lambda^x}{x!} = \\
&= \exp(-\lambda)\,\lambda \sum_{x=1}^{+\infty} x\,\frac{\lambda^{x-1}}{(x-1)!} = \\
&= \lambda \exp(-\lambda)\left(\sum_{x=1}^{+\infty}(x-1)\frac{\lambda^{x-1}}{(x-1)!} + \sum_{x=1}^{+\infty}\frac{\lambda^{x-1}}{(x-1)!}\right) = \\
&= \lambda \exp(-\lambda)\left(\lambda\sum_{x=1}^{+\infty}\frac{\lambda^{x-2}}{(x-2)!} + \sum_{x=1}^{+\infty}\frac{\lambda^{x-1}}{(x-1)!}\right) = \\
&= \lambda\,(\lambda+1) = \lambda^2 + \lambda
\end{aligned}
\tag{13}
$$

We can now evaluate:

$$
Var(X) = E[X^2] - (E[X])^2 = \lambda \tag{14}
$$

1.2 First observe that $E[T] = \lim_{N\to\infty}\sum_{n=1}^{N} n\,P(T=n)$. Then, since:

$$
\begin{aligned}
&\sum_{n=1}^{N} n\,P(T=n) = \\
&= P(T=1) + P(T=2) + P(T=2) + P(T=3) + P(T=3) + P(T=3) + \cdots = \\
&= (P(T=1) + \cdots + P(T=N)) + (P(T=2) + \cdots + P(T=N)) + \\
&+ \cdots + (P(T=N-1) + P(T=N)) + P(T=N) = \sum_{n=1}^{\infty} P(N \geq T \geq n)
\end{aligned}
\tag{15}
$$

one has:

$$
E[T] = \lim_{N\to\infty}\sum_{n=1}^{\infty} P(N \geq T \geq n) = \sum_{n=1}^{\infty} P(T \geq n) \tag{16}
$$

The second point is a simple application of multi-dimensional integration techniques.

$$
\begin{aligned}
E[X] &= \int_0^{+\infty} dx\, x\, p(x) = \int_0^{+\infty} dx\, p(x) \int_0^x dy = \\
&= \int_0^{+\infty} dx \int_0^x dy\, p(x) = \\
&= \int_0^{+\infty} dy \int_y^{+\infty} dx\, p(x) = \\
&= \int_0^{+\infty} dy\, P\,(X \geq y) = \int_0^{+\infty} dy\,(1 - F(y))
\end{aligned}
\tag{17}
$$

1.3 The angle θ between A and B is a random variable uniformly distributed on the interval $[-\pi, \pi]$. The length $L(\theta)$ of the chord connecting A and B is:

$$L(\theta) = \sqrt{2}\sqrt{1 - \cos(\theta)} \tag{18}$$

The cumulative distribution function is, for $l \in (0, 2)$:

$$F(l) = P(L(\theta) \le l) = P(-\theta_0 \le \theta \le \theta_0) = \frac{\theta_0}{\pi} = \frac{arccos(1 - l^2/2)}{\pi} \tag{19}$$

The probability density is $1 - F(\sqrt{3}) = \frac{1}{3}$.

$$p_L(l) = \frac{dF(l)}{dl} = \frac{1}{\pi}\frac{1}{\sqrt{1 - l^2/4}} \tag{20}$$

So that $E[L] = \int_0^2 dl \, l \, p_L(l) = \frac{4}{\pi}$ and $var(L) = E[L^2] - E[L]^2 = 2 - \frac{16}{\pi^2}$.

1.4   1. Suppose that, fixed a point A on the circumference a point x is chosen in the radius ending on A with uniform probability; the length of the chord passing through x and perpendicular to the radius ending on A is:

$$L(x) = 2\sqrt{1 - x^2} \tag{21}$$

   the chord is longer than $\sqrt{3}$ if $x \le \frac{1}{2}$, then $P = \frac{1}{2}$.
   2. Suppose that a point x with polar coordinates (r, θ) is chosen randomly anywhere within the circle having the circumference as its boundary; the length of the chord having such point as midpoint is $L(r, \theta) = 2\sqrt{1 - r^2}$, and it is longer than $\sqrt{3}$ if the point falls within a circle of radius $\frac{1}{2}$; since the area of such small circle is $\frac{\pi}{4}$ then $P = \frac{1}{4}$.

1.5 Let's face the calculation of:

$$I_0 = \int_{\mathbb{R}^d} d\mathbf{x} \, \exp\left(-\frac{1}{2}\sum_{i,j=1}^{d} x_i \mathcal{O}_{ij} x_j\right) \tag{22}$$

since the first part of the exercise is a simple corollary of this result.

Since the matrix $\mathcal{O}$ is real, symmetric and positive definite, it can be diagonalized through an orthogonal matrix $\mathcal{R}$:

$$\mathcal{R}^T \mathcal{O} \mathcal{R} = \begin{pmatrix} \lambda_1 & 0 & \dots & 0 \\ 0 & \lambda_2 & \dots & 0 \\ \dots & \dots & \dots & \dots \\ 0 & 0 & \dots & \lambda_d \end{pmatrix} \tag{23}$$

We then perform a change of variables inside the integral $\mathbf{x} = \mathcal{R}\mathbf{y}$ and keep in mind that the orthogonal matrix has unit jacobian. We obtain:

$$I_0 = \int_{\mathbb{R}^d} d\mathbf{y}\, \exp\left(-\frac{1}{2}\sum_{i=1}^d y_i\, \lambda_i\, y_i\right) = \prod_{i=1}^d \int_{-\infty}^{+\infty} dy_i\, \exp\left(-\frac{1}{2}\lambda_i y_i^2\right) \tag{24}$$

that is:

$$I_0 = \prod_{i=1}^d \sqrt{\frac{2\pi}{\lambda_i}} = \frac{(2\pi)^{d/2}}{\sqrt{\det(\mathcal{O})}} \tag{25}$$

We have thus found that:

$$\int_{\mathbb{R}^d} d\mathbf{x}\, \exp\left(-\frac{1}{2}\sum_{i,j=1}^d x_i \mathcal{O}_{ij} x_j\right) = \frac{(2\pi)^{d/2}}{\sqrt{\det(\mathcal{O})}} \tag{26}$$

Now let's turn to:

$$I = \int_{\mathbb{R}^d} d\mathbf{x}\, \exp\left(-\frac{1}{2}\sum_{i,j=1}^d x_i \mathcal{O}_{ij} x_j + \sum_{i=1}^d \vartheta_i x_i\right) \tag{27}$$

Using the same change of variables as before we find:

$$I = \int_{\mathbb{R}^d} d\mathbf{y}\, \exp\left(-\frac{1}{2}\sum_i^d \lambda_i y_i^2 + \sum_{i=1}^d \left(\mathcal{R}^T \vartheta\right)_i y_i\right) \tag{28}$$

that is:

$$I = \prod_{i=1}^d \int_{-\infty}^{+\infty} dy_i\, \exp\left(-\frac{1}{2}\lambda_i y_i^2 + \left(\mathcal{R}^T \vartheta\right)_i y_i\right) \tag{29}$$

If we write:

$$\frac{1}{2}\lambda_i y_i^2 - \left(\mathcal{R}^T \vartheta\right)_i y_i = \frac{\lambda_i}{2}\left(y_i - \frac{\left(\mathcal{R}^T \vartheta\right)_i}{\lambda_i}\right)^2 - \frac{1}{2}\frac{\left(\mathcal{R}^T \vartheta\right)_i^2}{\lambda_i} \tag{30}$$

and shift the integration variable we get:

$$I = \prod_{i=1}^d \exp\left(\frac{1}{2}\frac{\left(\mathcal{R}^T \vartheta\right)_i^2}{\lambda_i}\right) \int_{-\infty}^{+\infty} dy_i\, \exp\left(-\frac{1}{2}\lambda_i y_i^2\right) =$$

$$= \prod_{i=1}^d \exp\left(\frac{1}{2}\frac{\left(\mathcal{R}^T \vartheta\right)_i^2}{\lambda_i}\right)\sqrt{\frac{2\pi}{\lambda_i}} = \frac{(2\pi)^{d/2}}{\sqrt{\det(\mathcal{O})}}\exp\left(\sum_{i=1}^d \frac{1}{2}\frac{\left(\mathcal{R}^T \vartheta\right)_i^2}{\lambda_i}\right) \tag{31}$$

Let's conclude now observing that:

$$\sum_{i=1}^{d} \frac{\left(\mathcal{R}^T \vartheta\right)_i^2}{\lambda_i} = \sum_{i,j,k=1}^{d} \frac{\left(\mathcal{R}_{ij}^T \vartheta_j\right)\left(\mathcal{R}_{ik}^T \vartheta_k\right)}{\lambda_i} = \sum_{j,k=1}^{d} \vartheta_j \vartheta_k \sum_{i=1}^{d} \mathcal{R}_{ji} \frac{1}{\lambda_i} \mathcal{R}_{ik}^T \qquad (32)$$

and that:

$$\sum_{i=1}^{d} \mathcal{R}_{ji} \frac{1}{\lambda_i} \mathcal{R}_{ik}^T = \mathbb{O}_{jk}^{-1} \qquad (33)$$

We have thus the following expression:

$$I = \frac{(2\pi)^{d/2}}{\sqrt{\det(\mathbb{O})}} \exp\left(\frac{1}{2} \sum_{j,k=1}^{d} \vartheta_j \mathbb{O}_{jk}^{-1} \vartheta_k\right) \qquad (34)$$

It is thus immediate to obtain:

$$\frac{\partial^2 I}{\partial \vartheta_i \partial \vartheta_j}\big|_{\vartheta=0} = \mathbb{O}_{ij}^{-1} \frac{(2\pi)^{d/2}}{\sqrt{\det(\mathbb{O})}} \qquad (35)$$

On the other hand we observe that:

$$\frac{\partial^2 I}{\partial \vartheta_i \partial \vartheta_j}\big|_{\vartheta=0} = \int_{\mathbb{R}^d} d\mathbf{x}\, x_i x_j \exp\left(-\frac{1}{2} \sum_{i,j=1}^{d} x_i \mathbb{O}_{ij} x_j\right) \qquad (36)$$

Putting all together, we have the result:

$$\frac{\sqrt{\det(\mathbb{O})}}{(2\pi)^{d/2}} \int_{\mathbb{R}^d} d\mathbf{x}\, x_i x_j \exp\left(-\frac{1}{2} \sum_{i,j=1}^{d} x_i \mathbb{O}_{ij} x_j\right) = \mathbb{O}_{ij}^{-1} \qquad (37)$$

In the language of quantum field theory this result states that, in a free (gaussian) theory, the two point function $E[X_i X_j] \equiv \langle x_i x_j \rangle$ is equal to the inverse of the propagator (the matrix $\mathbb{O}$).

1.6 The characteristic function of a random variable $X \sim N(0, \sigma^2)$ is:

$$\phi_X(t) = E[e^{itX}] = e^{-\frac{E[X^2]t^2}{2}} \qquad (38)$$

Choosing $t = 1$ and expanding both members of the previous equality in series:

$$\sum_{r=0}^{\infty} \frac{(-1)^r}{(2r)!} E[X^{2r}] = \sum_{s=0}^{\infty} \frac{(-1)^s}{2^s s!} E[X^2]^s \qquad (39)$$

Applying the previous identity to $X = c_1 Z_1 + \cdots + c_{2n} Z_{2n}$, one obtains two analytic functions in the $c_1 \ldots c_{2n}$ coefficients. The left member reads:

$$\sum_{r=0}^{\infty} \frac{(-1)^r}{(2r)!} \sum_{k_1 \ldots k_{2r}=0}^{\infty} \binom{2r}{k_1 \ldots k_{2r}} c_1^{k_1} \ldots c_{2r}^{k_{2r}} E[Z_1^{r_1} \ldots Z_{2r}^{r_{2r}}] \tag{40}$$

with the constraint $k_1 + \cdots + k_{2r} = 2r$. The only term proportional to $c_1 \ldots c_{2n}$ in the above series corresponds to $r = n$, $k_1 = \cdots = k_{2r} = 1$:

$$\frac{(-1)^n}{(2n)!}(2n)! c_1 \ldots c_{2r} E[Z_1 \ldots Z_{2n}] \tag{41}$$

The right member of (39) reads:

$$\sum_{s=0}^{\infty} \frac{(-1)^s}{2^s s!} \left(\sum_{ij=1}^{2n} c_i c_j E[Z_i Z_j] \right)^s = \sum_{s=0}^{\infty} \frac{(-1)^s}{2^s s!} \sum_{i_1 j_1 \ldots i_s j_s = 1}^{2n} c_{i_1} c_{j_1} E[Z_{i_1} Z_{j_1}] \ldots c_{i_s} c_{j_s} E[Z_{i_s} Z_{j_s}] \tag{42}$$

The only terms proportional to $c_1 \ldots c_{2n}$ correspond to $s = n$, $i_1 j_1 \ldots i_s j_s$ permutation of $1 \ldots 2n$:

$$\frac{(-1)^n}{2^n n!} \sum_{\sigma \in S_{2n}} E[Z_{\sigma(1)} Z_{\sigma(2)}] \ldots E[Z_{\sigma(2n-1)} Z_{\sigma(2n)}] c_1 \ldots c_{2n} \tag{43}$$

Putting all together we find the identity:

$$E[Z_1 \ldots Z_{2n}] = \frac{1}{2^n n!} \sum_{\sigma \in S_{2n}} E[Z_{\sigma(1)\sigma(2)}] \ldots E[Z_{\sigma(2n-1)\sigma(2n)}] \tag{44}$$

The rearrangement necessary to write down Wick formula simply follows from the observation that several terms in the above summation are redundant.

1.7 For all $y \in [0, \infty)$ the probability $P(Y \le y)$ is given by:

$$P(Y \le y) = \int_{\{x: |x| \le \sqrt{y}\}} dx_1 \ldots dx_n \, p(x_1) \ldots p(x_n) \tag{45}$$

Moving to spherical coordinates:

$$P(Y \le y) = \Omega_n \int_0^{\sqrt{y}} dx \, x^{n-1} \frac{e^{-\frac{x^2}{2}}}{(2\pi)^{\frac{n}{2}}} \tag{46}$$

where $\Omega_n = \frac{2\pi^{\frac{n}{2}}}{\Gamma(\frac{n}{2})}$. Differentiation with respect to y yields the desired probability density:

$$p(y) = \frac{1}{2^{\frac{n}{2}} \Gamma\left(\frac{n}{2}\right)} y^{\frac{n}{2}-1} e^{-\frac{y}{2}} \tag{47}$$

Moments of Y can be computed using the formula:

$$E[Y^m] = \int_0^\infty dy \, y^m \, p(y) = \frac{1}{2^{\frac{n}{2}} \Gamma\left(\frac{n}{2}\right)} \int_0^\infty dy \, y^{\frac{n}{2}+m-1} e^{-\frac{y}{2}} = 2^m \frac{\Gamma\left(m+\frac{n}{2}\right)}{\Gamma\left(\frac{n}{2}\right)} \tag{48}$$

In particular, $E[Y] = n$ and $var(Y) = 2n$.

1.8 For all $t \geq 0$ the event $T \leq t$ coincides with $Y \geq \frac{n}{t^2} X^2$, which has probability:

$$P(T \leq t) = \int_{-\infty}^0 dx \int_0^\infty dy \, p_X(x) \, p_Y(y) + \int_0^\infty dx \int_{\frac{nx^2}{t^2}}^\infty dy \, p_X(x) \, p_Y(y) =$$

$$= \frac{1}{2} + \int_0^\infty dx \int_{\frac{nx^2}{t^2}}^\infty dy \, p_X(x) \, p_Y(y) \tag{49}$$

Derivation with respect to t yields:

$$p(t) = \frac{2n}{t^3} \int_0^\infty dx \, x^2 \, p_X(x) \, p_Y\left(\frac{nx^2}{t^2}\right) =$$

$$= \frac{2n}{t^3} \int_0^\infty dx \, x^2 \frac{e^{-\frac{x^2}{2}}}{\sqrt{2\pi}} \frac{\left(\frac{nx^2}{t^2}\right)^{\frac{n}{2}-1} e^{-\frac{nx^2}{2t^2}}}{2^{\frac{n}{2}} \Gamma\left(\frac{n}{2}\right)} = \tag{50}$$

$$= \frac{\sqrt{2}}{2^{\frac{n}{2}} t^3} \left(\frac{n}{t^2}\right)^{\frac{n}{2}-1} \frac{n}{\Gamma\left(\frac{1}{2}\right)\Gamma\left(\frac{n}{2}\right)} \int_0^\infty dx \, x^n \, e^{-\frac{x^2}{2}\left(1+\frac{n}{t^2}\right)}$$

and since $\int_0^\infty dx \, x^n \, e^{-\frac{x^2}{2}\left(1+\frac{n}{t^2}\right)} = \frac{\Gamma\left(\frac{n+1}{2}\right) 2^{\frac{n-1}{2}}}{\left(1+\frac{n}{t^2}\right)^{\frac{n+1}{2}}}$ one readily obtains:

$$p(t) = \frac{1}{\sqrt{n}} \frac{1}{B\left(\frac{1}{2}, \frac{n}{2}\right)} \left(1 + \frac{t^2}{n}\right)^{-\frac{n+1}{2}} \tag{51}$$

Since T is an even random variable, its odd moments vanish. Moreover:

$$E(T^{2k}) = \int_{-\infty}^\infty dt \frac{t^{2k}}{\sqrt{n} B\left(\frac{1}{2}, \frac{n}{2}\right)} \left(1 + \frac{n}{t^2}\right)^{\frac{n+1}{2}} \tag{52}$$

this integral converges if and only if $k < \frac{n+1}{2}$. Were that the case, the change of coordinate $r = \frac{t^2}{n(1+\frac{t^2}{n})}$ yields:

$$E(T^{2k}) = \frac{n^k}{B\left(\frac{1}{2},\frac{n}{2}\right)} \int_0^1 dr\,(1-r)^{\frac{n}{2}-k-1} r^{k-\frac{1}{2}} = \frac{n^k}{B\left(\frac{1}{2},\frac{n}{2}\right)} B\left(k+\frac{1}{2},\frac{n}{2}-k\right)$$

(53)

In particular, $var(T) = E[T^2] = \frac{n}{n-2}$ provided that $n > 2$.

Problems of Chap. 2

2.1

$$P_\theta(\mathcal{T} \le t) = P_\theta(X_1 \le t, \ldots, X_n \le t) = \frac{t^n}{\theta^n} \chi_{[0,\theta]}(t)$$

(54)

therefore the statistics $\mathcal{T}$ has distribution:

$$p_\theta(t) = \frac{nt^{n-1}}{\theta^n} \chi_{[0,\theta]}(t)$$

(55)

As a consequence:

$$E_\theta[\mathcal{T}] = \int_0^\theta t\,\frac{nt^{n-1}}{\theta^n} = \frac{n}{n+1}\theta$$

(56)

$\mathcal{T}$ and the estimator is only asymptotically unbiased. Moreover, since:

$$E_\theta[\mathcal{T}^2] = \int_0^\theta t^2 \frac{nt^{n-1}}{\theta^n} = \frac{n}{n+2}\theta^2$$

(57)

we have:

$$var(\mathcal{T}) = \frac{n}{(n+1)(n+2)}\theta^2$$

(58)

so that $\mathcal{T}$ is consistent. Notice that $\mathcal{T}$ remains consistent even when multiplied by the factor $\frac{n+1}{n}$, which turns it into an unbiased estimator.

2.2

$$MSE(\mathcal{T}) = E\left[(\mathcal{T} - \tau(\theta))^2\right] = E\left[\mathcal{T}^2\right] + \tau(\theta)^2 - 2\tau(\theta)E[\mathcal{T}]$$

(59)

Writing $E\left[(\mathcal{T} - \tau(\theta))^2\right] = Var[\mathcal{T}] + E[\mathcal{T}^2]$ we find

$$MSE(\mathcal{T}) = Var[\mathcal{T}] + (\tau(\theta) - E[\mathcal{T}])^2$$

(60)

The bias $\tau(\theta) - E[\mathcal{T}]$ increases $MSE(\mathcal{T})$ beyond the minimum value $Var[\mathcal{T}]$.

2.3 Let's apply the Cochran theorem to the statistics $\mathcal{M}$ and $\mathcal{S}^2$. We let:

$$\tilde{e}_1 = \begin{pmatrix} \frac{1}{\sqrt{n}} \\ \cdots \\ \frac{1}{\sqrt{n}} \end{pmatrix}$$

and define $\tilde{E}_1$ as the one-dimensional subspace of $\mathbb{R}^n$ spanned by $\tilde{e}_1$. Moreover, we let $\tilde{E}_2$ be the orthogonal complement of $\tilde{E}_1$. Starting from the sample $X = (X_1, \ldots, X_n)$, we define the n-dimensional random variable:

$$Y = \frac{X - \theta_0 \sqrt{n}\, \tilde{e}_1}{\sqrt{\theta_1}} \tag{61}$$

which, by construction, follows the law $Y \sim N(\mathbf{0}, \mathbb{I})$. Using the notations of Cochran theorem, we have:

$$\Pi_1\, Y = \frac{\mathcal{M} - \theta_0}{\sqrt{\theta_1/n}}\, \tilde{e}_1 \qquad \Pi_2\, Y = \frac{X - \sqrt{n}\,\mathcal{M}\, \tilde{e}_1}{\sqrt{\theta_1}}$$

We know that $\Pi_1\, Y$ and $\Pi_2\, Y$ are **independent**. Moreover:

$$|\Pi_2\, Y|^2 = \frac{1}{\theta_1} \sum_{i=1}^{n} (X_i - \mathcal{M})^2 = \frac{(n-1)\mathcal{S}^2}{\theta_1} \tag{62}$$

has a law $\chi^2(n-1)$ and is **independent** on $\Pi_1\, Y$. It follows (see problem 1.7) that the random variable:

$$\mathcal{R} = \frac{\left(\frac{\mathcal{M}-\theta_0}{\sqrt{\theta_1}/\sqrt{n}} \right)}{\sqrt{|\Pi_2\, Y|^2/(n-1)}} = \frac{\mathcal{M} - \theta_0}{\sqrt{\mathcal{S}^2/n}} \tag{63}$$

has a **Student law** $\mathcal{R} \sim t(n-1)$ with $n-1$ degrees of freedom.

2.4 Keeping in mind the idea of dealing with samples with arbitrary size, We consider the sequence of r-dimensional random variables:

$$Z_n = \sqrt{n}\left(\mathcal{N} - p(\boldsymbol{\theta}_0) \right)$$

where:

$$\mathcal{N} = \begin{pmatrix} n_1(X_1 \ldots X_n) \\ \cdots \\ n_r(X_1 \ldots X_n) \end{pmatrix} \qquad p(\boldsymbol{\theta}_0) = \begin{pmatrix} p_1(\boldsymbol{\theta}_0) \\ \cdots \\ p_r(\boldsymbol{\theta}_0) \end{pmatrix}$$

The definition:

$$\mathscr{N} = \frac{1}{n} \sum_{i=1}^{n} \begin{pmatrix} 1_{E_1}(X_i) \\ \dots \\ 1_{E_r}(X_i) \end{pmatrix}$$

guarantees that $\mathscr{N}$ is the empirical mean of n random variables independent and identically distributed with mean:

$$E_\theta \left(\begin{pmatrix} 1_{E_1}(X_i) \\ \dots \\ 1_{E_r}(X_i) \end{pmatrix} \right) = p(\theta_0)$$

and covariance matrix:

$$\Sigma_{jk} = E_{\theta_0} \left(1_{E_j}(X_i) 1_{E_k}(X_i) \right) - E_{\theta_0} \left(1_{E_j}(X_i) \right) E_{\theta_0} \left(1_{E_k}(X_i) \right) =$$
$$= \delta_{jk}\, p_j(\theta_0) - p_j(\theta_0)\, p_k(\theta_0)$$

The multidimensional central limit theorem guarantees thus that:

$$\lim_{n\to\infty} Z_n = Z \sim N(\mathbf{0}, \Sigma)$$

where the convergence is meant in law. We observe now that the Pearson random variable may be expressed as an inner product in $\mathbb{R}^r$:

$$\mathscr{P} = (AZ_n | AZ_n)$$

where $A \in M_{r \times r}(\mathbb{R})$ is the diagonal matrix:

$$A = \text{diag} \left(\frac{1}{\sqrt{p_1(\theta_0)}} \dots \frac{1}{\sqrt{p_r(\theta_0)}} \right) \qquad (64)$$

Using the identities:

$$E_{\theta_0}(AZ_n) = A E_\theta(Z_n) = \mathbf{0} \qquad \text{Cov}(AZ_n) = A\,\text{Cov}(Z_n)\,A^T$$

we conclude that:

$$\lim_{n\to\infty} A Z_n = N(\mathbf{0}, A\Sigma A^T)$$

Performing the product of matrices we get:

$$(A\Sigma A^T)_{jk} = \delta_{jk} - \sqrt{p_j(\theta_0)}\sqrt{p_k(\theta_0)}$$

By inspection we see that $A \Sigma A^T$ is a projection matrix of rank:

$$\text{rg}(A \Sigma A^T) = \text{tr}(A \Sigma A^T) = \sum_k (A \Sigma A^T)_{kk} = r - 1$$

There exists thus an orthogonal matrix $U \in M_{r \times r}(\mathbb{R})$ such that:

$$A \Sigma A^T = U \, \text{diag}(1 \ldots 10) \, U^T \equiv U \, \Delta \, U^T$$

The sequence $U A Z_n$ converges thus in distribution to a random variable $N(\mathbf{0}, \Delta)$, whose components are independent: the first $r - 1$ follow a standard normal, the last one is the constant 0.

The sequence $(U A Z_n | U A Z_n)$ converges thus in law to a $\chi^2(r - 1)$ and:

$$\lim_{n \to \infty} \mathscr{P} = \lim_{n \to \infty} (A Z_n | A Z_n) = \lim_{n \to \infty} (U A Z_n | U A Z_n) \sim \chi^2(r - 1) \qquad (65)$$

This completes the proof.

2.5

$$\mathscr{M} = 852.4 \qquad (66)$$

$$\mathscr{S}^2 = 6242.7 \qquad (67)$$

A level 0.95 confidence interval for the mean is:

$$[\mathscr{A}, \mathscr{B}] = [836.7, 868.0] \qquad (68)$$

$c = 299852.4 \pm 15.67\,\text{km/s}$.

Stephen M Stigler, Ann. Stat. 5, 1055 (1977)

2.6 The observed numbers of samples in the bins $B_i = [(i - 1)\Delta x, i \Delta x)$, $i = 1 \ldots 5$, $\Delta x = \frac{1}{5}$, are $O_i = 3, 4, 2, 4, 7$ respectively. The expected numbers are $E_i \equiv 4$. So $\chi^2 = \sum_{i=1}^5 \frac{(O_i - E_i)^2}{E_i} = 3.5$. For $r = 5$, the $1 - \alpha = 0.9$ critical value is $\chi_{1-\alpha}(r - 1) = 7.78$. Since $\chi^2 < \chi_{1-\alpha}(r - 1)$ the hypothesis that $X_i \sim U(0, 1)$ is accepted with confidence level 0.9.

The maximum deviation $D = \sqrt{n} \sup_{[0,1]} |\tilde{F}_n(x) - F(x)|$ between the empirical

$$F_n(x) = \frac{1}{n} \sum_{i=1}^n \chi_{(-\infty, x]}(x_i) \qquad (69)$$

and exact $F(x) = x$ cumulative distribution function is $D = 0.9140$. In the limit of large sample (that we assume for simplicity), the $1 - \alpha = 0.9$ critical value is $D_{1-\alpha} = 1.222$. Since $D < D_{1-\alpha}$, the hypothesis that $X_i \sim U(0, 1)$ is accepted with confidence level 0.9.

2.7 Consider the statistics:

$$\mathcal{M}_X + \mathcal{M}_Y \quad \text{and} \quad \mathcal{M}_X \mathcal{M}_Y \tag{70}$$

since $\mathcal{M}_X \xrightarrow[n\to\infty]{a.s.} \mu_X$ and $\mathcal{M}_X \xrightarrow[n\to\infty]{a.s.} \mu_Y$, due to the continuous mapping theorem they represent asymptotically unbiased and consistent estimators for $\mu_X + \mu_Y$ and $\mu_X \mu_Y$.

For the purpose of constructing confidence intervals for $\mu_X + \mu_Y$, we observe that:

$$\sqrt{n}\,(\mathcal{M}_X + \mathcal{M}_Y - \mu_X - \mu_Y) \xrightarrow[n\to\infty]{L} N(0, \sigma_X^2 + \sigma_Y^2) \tag{71}$$

and that since $\mathscr{S}_X^2 + \mathscr{S}_Y^2 \xrightarrow[n\to\infty]{a.s.} \sigma_X^2 + \sigma_Y^2$, by virtue of Slutsky's Theorem:

$$\sqrt{n}\,\frac{(\mathcal{M}_X + \mathcal{M}_Y - \mu_X - \mu_Y)}{\sqrt{\mathscr{S}_X^2 + \mathscr{S}_Y^2}} \xrightarrow[n\to\infty]{L} N(0, 1) \tag{72}$$

Therefore $\left[\mathcal{M}_X + \mathcal{M}_Y - \dfrac{\sqrt{\mathscr{S}_X^2 + \mathscr{S}_Y^2}}{\sqrt{n}}\phi_{1-\frac{\alpha}{2}}, \mathcal{M}_X + \mathcal{M}_Y + \dfrac{\sqrt{\mathscr{S}_X^2 + \mathscr{S}_Y^2}}{\sqrt{n}}\phi_{1-\frac{\alpha}{2}} \right]$ is an approximate confidence interval for $\mu_X + \mu_Y$ of level $1 - \alpha$.

Analogous calculations show that:

$$\sqrt{n}\,(\mathcal{M}_X \mathcal{M}_Y - \mu_X \mu_Y) \xrightarrow[n\to\infty]{L} N\left(0, (\mu_X \mu_Y)^2 \left(\frac{\sigma_X^2}{\mu_X^2} + \frac{\sigma_Y^2}{\mu_Y^2}\right)\right) \tag{73}$$

and that $\left[\mathcal{M}_X \mathcal{M}_Y - \dfrac{\mathcal{M}_X \mathcal{M}_Y \sqrt{\frac{\mathscr{S}_X^2}{\mathcal{M}_X^2} + \frac{\mathscr{S}_Y^2}{\mathcal{M}_Y^2}}}{\sqrt{n}}\phi_{1-\frac{\alpha}{2}}, \mathcal{M}_X \mathcal{M}_Y + \dfrac{\mathcal{M}_X \mathcal{M}_Y \sqrt{\frac{\mathscr{S}_X^2}{\mathcal{M}_X^2} + \frac{\mathscr{S}_Y^2}{\mathcal{M}_Y^2}}}{\sqrt{n}}\phi_{1-\frac{\alpha}{2}} \right]$ is

an approximate confidence interval for $\mu_X \mu_Y$ of level $1 - \alpha$.

Problems of Chap. 3

3.1 We proceed as follows:

$$P(S_N = k) = P\left(\bigcup_{n \in \mathbb{N}} (\{S_N = k\} \cap \{N = n\})\right) = \sum_{n \in \mathbb{N}} P(S_n = k, N = n) \tag{74}$$

where we used the fact that if $N = n$ then $S_N = S_n$. Now, using independence, we have:

$$P(S_N = k) = \sum_{n \in \mathbb{N}} P(S_n = k)\, P(N = n) \tag{75}$$

We have:

$$\psi_{S_N}(z) = E[z^{S_N}] = \sum_k z^k P(S_N = k) = \sum_{k,n} z^k P(S_n = k) P(N = n) = \sum_n P(N = n)\psi_{S_n}(z)$$
(76)

We have also:
$$\psi_{S_n}(z) = \left(\psi_{X_i}(z)\right)^n$$
(77)

so that:
$$\psi_{S_N}(z) = \sum_n P(N = n) \left(\psi_{X_i}(z)\right)^n = \psi_N \left(\psi_{X_i}(z)\right)$$
(78)

Finally, differentiating, we have:

$$E[S_N] = \psi_{S_N}^{(1)}(z = 1) = \psi_N^{(1)} \left(\psi_{X_i}(z = 1)\right) \psi_{X_i}^{(1)}(z = 1) = E[N]E[X_i]$$
(79)

since any generating function is equal to 1 if $z = 1$.

3.2 We compute:

$$P(X \geq j + k \,|\, X \geq j) = \frac{P(X \geq j + k, X \geq j)}{P(X \geq j)}$$
$$= \frac{P(X \geq j + k)}{P(X \geq j)}$$
(80)

Now, the identity:

$$P(X \geq j) = \sum_{x=j}^{+\infty} p(1 - p)^x = (1 - p)^j$$
(81)

implies that:

$$P(X \geq j + k \,|\, X \geq j) = \frac{P(X \geq j + k)}{P(X \geq j)} = \frac{(1 - p)^{j+k}}{(1 - p)^j} = (1 - p)^k = P(X \geq k)$$
(82)

In the case of the exponential distribution, the same result can be obtained following the same procedure, using the fact that:

$$P(X \geq t) = \int_t^{+\infty} dx \, \lambda e^{-\lambda x} = e^{-\lambda t}$$
(83)

3.3 We denote M the event that the individual has the disease and A the event that the test turns out positive. We have, $P(A|M) = P(A|M^c) = 0.99$, $P(M) = 0.002$ We apply Bayes theorem to find:

$$P(M|A) = \frac{P(A|M)P(M)}{P(A)} = \frac{P(A|M)P(M)}{P(A|M)P(M) + P(A|M^c)(1 - P(M))} \quad (84)$$

3.4 The result is immediately proved observing that, if we let:

$$W = \begin{cases} \frac{1}{P(A)} \int_A P(d\omega)X(\omega), & \omega \in A \\ \frac{1}{P(A^c)} \int_{A^c} P(d\omega)X(\omega), & \omega \in A^C \end{cases} \quad (85)$$

then:

$$E[1_A W] = P(A) \times \frac{1}{P(A)} \int_A P(d\omega)X(\omega) = E[1_A X] \quad (86)$$

and the same holds for $A^C, \emptyset, \Omega$. This implies that:

$$W = E[X|\mathscr{G}] \quad (87)$$

being manifestly $\mathscr{G}$-measurable.

3.5 Let us start computing the marginal probability density:

$$p_Y(y) = \begin{cases} 0, & |y| > 1 \\ \frac{1}{\pi} \int_{-\sqrt{1-y^2}}^{\sqrt{1-y^2}} dx, & |y| \le 1 \end{cases} = \frac{2}{\pi}\sqrt{1 - y^2}1_{(-1,1)}(y) \quad (88)$$

We have thus:

$$p(x|y) = \frac{p_{(X,Y)}(x, y)}{p_Y(y)} = \frac{1}{2}\frac{1}{\sqrt{1 - y^2}}1_{\{x^2+y^2<1\}}(x, y) \quad (89)$$

We know that:

$$E[X|Y] = g(Y), \quad g(y) = \int_{-\infty}^{+\infty} dx\, x\, p(x|y) = 0 \quad (90)$$

3.6 We can model the situation as follows:

$$p(x|m) = \frac{1}{\sqrt{2\pi}}e^{-\frac{1}{2}(x-m)^2} \quad (91)$$

while:

$$p(m) = \lambda e^{-\lambda m} 1_{(0,+\infty)}(m) \quad (92)$$

We observe that:

$$\int_{-\infty}^{+\infty} dx\, x\, p(x|m) = m \quad (93)$$

which means that $E[X|M] = M$, whence we find:

$$E[X] = E[E[X|M]] = E[M] = \frac{1}{\lambda} \tag{94}$$

We finally have:

$$p(x, m) = p(x|m)p(m) = \frac{\lambda}{\sqrt{2\pi}} e^{-\frac{1}{2}(x-m)^2 - \lambda m} 1_{(0,+\infty)}(m) \tag{95}$$

3.7 We have:

$$p(n|\lambda) = \frac{\lambda^n}{n!} e^{-\lambda} \tag{96}$$

We have thus:

$$\begin{aligned}
p_N(n) &= \int_0^{+\infty} d\lambda \, \frac{\lambda^n}{n!} e^{-\lambda} \frac{\beta^\alpha}{\Gamma(\alpha)} \lambda^{\alpha-1} \exp(-\beta\lambda) \\
&= \frac{1}{\Gamma(\alpha) \, n!} \beta^\alpha \int_0^{+\infty} d\lambda \, \lambda^{n+\alpha-1} \exp(-(\beta+1)\lambda) \\
&\frac{1}{\Gamma(\alpha) \, n!} \beta^\alpha \frac{1}{(\beta+1)^{n+\alpha}} \int_0^{+\infty} dx \, x^{n+\alpha-1} \exp(-x) = \frac{\Gamma(n+\alpha)}{\Gamma(\alpha) \, n!} \beta^\alpha \frac{1}{(\beta+1)^{n+\alpha}}
\end{aligned} \tag{97}$$

which is the desired result.

A simple calculation shows that:

$$E[N] = \alpha \frac{1-p}{p} \tag{98}$$

Problems of Chap. 4

4.1 The two random variables X_n, number of molecules in the first container at time instant n, and Y_n, number of molecules in the second container at time instant n, are related by the constraint $X_n + Y_n = N$. The law of X_0 is $(1, 0, \dots, 0)$.

This chain is irreducible, since all the states communicate. Studying the equation $\pi P = \pi$:

$$\pi_0 = q_1 \pi_1$$
$$\pi_1 = p_0 \pi_0 + q_2 \pi_2$$
$$\cdots$$
$$\pi_k = \pi_{k-1} p_{k-1} + \pi_{k+1} p_{k+1}$$
$$\cdots$$
$$\pi_N = \pi_{N-1} p_{N-1}$$

we find a unique invariant law, whose entries satisfy the following:

$$\pi_i = \frac{p_{i-1} \cdots p_0}{q_i \cdots q_1} \pi_0 = \binom{N}{i} \pi_0$$

Observing that $\sum_i \pi_i = 2^N \pi_0$ we get $\pi_0 = 2^{-N}$, so that we have finally:

$$\pi_i = \binom{N}{i} 2^{-N}$$

The expectation of the law of the random variable whose density is π is $N/2$, while its variance is $N/4$.

4.2 A general property of transition matrices ensures us that $(1, 1, 1)$ is a right eigenvalue of the matrix P; we observe that it is also a left eigenvalue. The similarity transformation $U^T P U$ with the orthogonal matrix:

$$U = \begin{pmatrix} \frac{1}{\sqrt{3}} & -\frac{1}{\sqrt{6}} & -\frac{1}{\sqrt{2}} \\ \frac{1}{\sqrt{3}} & \frac{2}{\sqrt{6}} & 0 \\ \frac{1}{\sqrt{3}} & -\frac{1}{\sqrt{6}} & \frac{1}{\sqrt{2}} \end{pmatrix}$$

leads to:

$$U^T P U = \begin{pmatrix} 1 & 0 & 0 \\ 0 & -\frac{1}{2} & \frac{\sqrt{3}}{2}(2p-1) \\ 0 & -\frac{\sqrt{3}}{2}(2p-1) & -\frac{1}{2} \end{pmatrix}$$

The eigenvalues of the 2×2 block are $-\frac{1}{2} \pm \frac{\sqrt{3}}{2} i \sqrt{4p^2 + 1 - 4p}$ and their modulus is $\sqrt{1 + 3p^2 - 3p}$, strictly less than 1 if $p \in (0, 1)$.

If $p \in (0, 1)$ the n-th power of the modulus of those eigenvalues tends to zero, and thus:

$$U^T P^n U \to \begin{pmatrix} 1 & 0 & 0 \\ 0 & 0 & 0 \\ 0 & 0 & 0 \end{pmatrix} \iff P^n \to \frac{1}{3}\begin{pmatrix} 1 & 1 & 1 \\ 1 & 1 & 1 \\ 1 & 1 & 1 \end{pmatrix}$$

implying that P has a unique stationary distribution, $(\frac{1}{3}, \frac{1}{3}, \frac{1}{3})$. If $p = 0$ or $p = 1$, the random walk is simply a rotation, clockwise or counterclockwise. The eigenvalues of P in such case are the three complex square roots of 1. The limit of the matrix P for $n \to \infty$ does not exist, and there are no invariant laws.

4.3 The first calculation is very simple

$$E[Z_{n+1} | Z_n = k] = E[\xi_1^n + \cdots + \xi_k^n] = k\mu$$

where μ is the expectation of ξ_k^n, assumed independent from n and k. Then $E[Z_{n+1} | Z_n] = \mu Z_n$.

Similarly:

$$E[Z_{n+1}^2|Z_n = k] = E[(\xi_1^n + \cdots + \xi_k^n)^2] = k^2\mu^2 + k\sigma$$

where σ is the variance of ξ_k^n.

We have thus $E[Z_{n+1}^2|Z_n] = \mu^2 Z_n^2 + \sigma Z_n$.

Using properties of conditional expectation we conclude that:

$$E[Z_{n+1}] = E[E[Z_{n+1}|Z_n]] = \mu E[Z_n]$$

$$E[Z_{n+1}^2] = E[E[Z_{n+1}^2|Z_n]] = \mu^2 E[Z_n^2] + \sigma \mu^n$$

$$Var[Z_{n+1}] = \mu^2 Var[Z_n] + \sigma \mu^n$$

The recursion relation for the expectations implies that:

$$E[Z_n] = \mu^n \tag{99}$$

while the one for the variances is satisfied by

$$Var[Z_n] = \begin{cases} \sigma \dfrac{\mu^n}{\mu} \dfrac{1-\mu^n}{1-\mu}, & \mu \neq 1 \\ n\sigma, & \mu = 1 \end{cases} \tag{100}$$

If $\mu < 1$ the average of the population goes to zero, together with its variance. If $\mu = 1$ ($\mu > 1$) the average of the population remains constant (diverges).

Now let's turn to the extinction problem. We will need a few definitions:

$$\varphi(t) = E\left[t^{\xi_i^{(n)}}\right] = \sum_{k=0}^{+\infty} p_k t^k \tag{101}$$

and:

$$\varphi_n(t) = E\left[t^{Z_n}\right] = \sum_{k=0}^{+\infty} P(Z_n = k) t^k \tag{102}$$

The following is very important:

$$\varphi_{n+1}(t) = E\left[t^{Z_{n+1}}\right] = \sum_{k=0}^{+\infty} E\left[t^{Z_{n+1}}|Z_n = k\right] P(Z_n = k)$$

$$= \sum_{k=0}^{+\infty} E\left[t^{\sum_{i=0}^{k} \xi_i^{(n)}}\right] P(Z_n = k) \tag{103}$$

$$= \sum_{k=0}^{+\infty} E\left[t^{\xi_i^{(n)}}\right]^k P(Z_n = k) = \varphi_n\left(\varphi(t)\right)$$

We will prove now by induction that also:

$$\varphi_{n+1}(t) = \varphi(\varphi_n(t)) \tag{104}$$

Since $Z_0 = 1$, $\varphi_0(t) = t$ so that the above is verified if $n = 0$. Now we assume (104) true for n, and we observe that:

$$\varphi_{n+2}(t) = \varphi_{n+1}(\varphi(t)) = \varphi(\varphi_n(\varphi(t))) = \varphi(\varphi_{n+1}(t)) \tag{105}$$

where we used twice (103).

Also:

$$\varphi_n(0) \overset{def}{=} \lim_{t \to 0} \varphi_n(t) = \lim_{t \to 0} \sum_{k=0}^{+\infty} P(Z_n = k) t^k = P(Z_n = 0) \tag{106}$$

We observe now that:

$$\mathscr{E} = \bigcup_{n \geq 1} \{Z_n = 0\} \tag{107}$$

Since the events are encapsulated, that is $\{Z_n = 0\} \subset \{Z_{n+1} = 0\}$, we have, for the extinction probability:

$$p_{ext} \overset{def}{=} P(\mathscr{E}) = \lim_{n \to +\infty} P(Z_n = 0) = \lim_{n \to +\infty} \varphi_n(0) \tag{108}$$

Now we observe that:

$$\varphi(p_{ext}) = \varphi(\lim_{n \to +\infty} \varphi_n(0)) = \lim_{n \to +\infty} \varphi(\varphi_n(0))$$
$$= \lim_{n \to +\infty} \varphi_n(\varphi(0)) = \lim_{n \to +\infty} \varphi_{n+1}(0) = p_{ext} \tag{109}$$

We have thus found that p_{ext} satisfies the fixed-point equation:

$$p_{ext} = \varphi(p_{ext}) = \sum_{k=0}^{+\infty} p_k p_{ext}^k \tag{110}$$

For Lotka distribution, we get $p_{ext} \sim 0.598271$.

4.4 Denote the probability that the first gambler is ruined having started with a coins through $P(R_a)$, that is:

$$R_a = \{\omega \in \Omega \mid \lim_{n \to +\infty} X_n(\omega) = 0, X_0 = a\} \tag{111}$$

X_n being the amount of coins after n rounds.

Obviously $P(R_0) = 1$ and $P(R_N) = 0$; otherwise:

$$P(R_a) = P(R_a|X_1 = a+1)p(X_1 = a+1) + P(R_a|X_1 = a-1)p(X_1 = a-1) = P(R_{a+1})p + P(R_{a-1})q \tag{112}$$

it is easily shown that such recursion relation is solved, together with the boundary conditions $P(R_0) = 1$ and $P(R_N) = 0$, by:

$$P(R_a) = \frac{N-a}{N} \quad \text{if } p = 1/2$$

$$P(R_n) = \frac{\left(\frac{p}{q}\right)^a - \left(\frac{p}{q}\right)^N}{1 - \left(\frac{p}{q}\right)^N} \quad \text{otherwise} \tag{113}$$

Observe that if $\frac{p}{q} \to 1$ the probability $P(R_a)$ converges, by l'Hopital's Rule, to $\frac{N-a}{N}$.

4.5 This is a simple algebraic exercise.

4.6
Let the state space of the random walk be the set $E = \{1, 2, 3, 4, 5, 6, 7, 8, 9\}$. Since a room i with i even is connected only to rooms j with j odd and viceversa (for instance, room 2 is connected to rooms 1, 3, 5 and thus $P_{2 \to j} = \frac{1}{3}$ for $j = 1, 3, 5$ and 0 otherwise). It is therefore convenient to order the elements of E separating odd and even numbers, for instance $E = \{1, 3, 7, 9, 5, 2, 4, 6, 8\}$ and writing the transition matrix of the process as:

$$P_{ij} = P_{i \to j} = \begin{pmatrix} 0 & 0 & 0 & 0 & 0 & 1/2 & 1/2 & 0 & 0 \\ 0 & 0 & 0 & 0 & 0 & 1/2 & 0 & 1/2 & 0 \\ 0 & 0 & 0 & 0 & 0 & 0 & 1/2 & 0 & 1/2 \\ 0 & 0 & 0 & 0 & 0 & 0 & 0 & 1/2 & 1/2 \\ 0 & 0 & 0 & 0 & 0 & 1/4 & 1/4 & 1/4 & 1/4 \\ 1/3 & 1/3 & 0 & 0 & 1/3 & 0 & 0 & 0 & 0 \\ 1/3 & 0 & 1/3 & 0 & 1/3 & 0 & 0 & 0 & 0 \\ 0 & 1/3 & 0 & 1/3 & 1/3 & 0 & 0 & 0 & 0 \\ 0 & 0 & 1/3 & 1/3 & 1/3 & 0 & 0 & 0 & 0 \end{pmatrix} \tag{114}$$

for instance, the only nonzero elements of the $i = 2$ row correspond to the columns $j = 1, 3, 5$. If the monkey is in a state i with i even (odd), it can return to i in at least 2 steps. This matrix is irreducible (in fact any two rooms i, j communicate). To find the invariant distribution, we solve:

$$\pi = \pi \, P_{ij} \tag{115}$$

and find:

$$\pi = \frac{1}{24} \, (2, 2, 2, 2, 4, 3, 3, 3) \tag{116}$$

We observe that this invariant distribution carries an intuitive meaning: the element i is proportional to the number of doors leading to the cell i.

4.7 This exercise allows to discuss an interesting feature of Markov chains, that is the concepts of transient and absorbing states. We start at a more general level.

Suppose a Markov chain has state space E and transition probability P such that $P_{a \to i} = 0$ for all $i < k$, $a \geq k$ and $P_{a \to b} = \delta_{a,b}$ for all $a, b \geq k$ for a given integer k. The states $a, b \geq k$ are called absorbing. All the other states $i, j < k$ are called transient and we will keep consistent notations throughout this solution.

The transition matrix of the chain thus takes the form

$$P = \begin{pmatrix} Q & R \\ 0 & I \end{pmatrix} \tag{117}$$

where Q, R describe transitions between transient states and from transient to absorbing states respectively. We assume that R is a non-zero matrix.

With a simple proof by induction over $n \geq 1$,

$$P^n = \begin{pmatrix} Q^n & \left(\sum_{k=0}^{n} Q^k \right) R \\ 0 & I \end{pmatrix} \tag{118}$$

The entry $(Q^n)_{ij}$ of Q^n gives the probability for being in a transient state j after n step, starting from a transient state i.

In the limit $n \to \infty$, provided that the limit exists, one has

$$\lim_{n \to +\infty} P^n = \begin{pmatrix} 0 & (1 - Q)^{-1} R \\ 0 & I \end{pmatrix} \tag{119}$$

The ia-entry of $(1 - Q)^{-1} R$ gives the probability that chain is absorbed to a state a, having started from the state i.

Assuming moreover $X_0 = i$, the expected number of times the chain is in the transient state j reads:

$$N_{ij} = E \left[\sum_{n=0}^{+\infty} 1_{\{X_n = j\}} \right] = \sum_{n=0}^{+\infty} (Q^n)_{ij} = (1 - Q)^{-1}_{ij} \tag{120}$$

The expected total number t_i of steps before the chain is absorbed, having started from i, is therefore

$$t_i = \sum_j N_{ij} = \sum_j (1 - Q)_{ij}^{-1} \quad . \tag{121}$$

This quantity is the average duration of the absorption process, assuming the chain started in i.

Now we come back to the monkey. We add a cell, say 10, accessible only starting from cell 9:

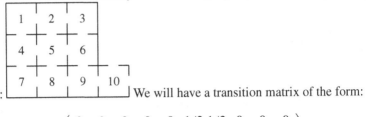

We will have a transition matrix of the form:

$$P_{ij} = P_{i \to j} = \begin{pmatrix} 0 & 0 & 0 & 0 & 0 & 1/2 & 1/2 & 0 & 0 & 0 \\ 0 & 0 & 0 & 0 & 0 & 1/2 & 0 & 1/2 & 0 & 0 \\ 0 & 0 & 0 & 0 & 0 & 0 & 1/2 & 0 & 1/2 & 0 \\ 0 & 0 & 0 & 0 & 0 & 0 & 0 & 1/3 & 1/3 & 1/3 \\ 0 & 0 & 0 & 0 & 0 & 1/4 & 1/4 & 1/4 & 1/4 & 0 \\ 1/3 & 1/3 & 0 & 0 & 1/3 & 0 & 0 & 0 & 0 & 0 \\ 1/3 & 0 & 1/3 & 0 & 1/3 & 0 & 0 & 0 & 0 & 0 \\ 0 & 1/3 & 0 & 1/3 & 1/3 & 0 & 0 & 0 & 0 & 0 \\ 0 & 0 & 1/3 & 1/3 & 1/3 & 0 & 0 & 0 & 0 & 0 \\ 0 & 0 & 0 & 0 & 0 & 0 & 0 & 0 & 0 & 1 \end{pmatrix} \tag{122}$$

where the new state space is $E' = \{1, 3, 7, 9, 5, 2, 4, 6, 8, 10\}$. Assuming that the monkey starts in cell 1, using formula (121), we obtain an average of 43 steps.

Problems of Chap. 5

5.1 For simplicity, without loosing generality, we consider only two parallel lines at distance d and we allow the center of mass of the needle to fall at any point within the area included between the lines. Let Θ be a random variable representing the slope of the needle; more precisely, Θ will be the angle formed by the needle measured with respect to the lines. Considering symmetry, we can assume that Θ follows a uniform law on $[0, \frac{\pi}{2}]$. We introduce another random variable X, representing the distance between the center of mass of the needle and one straight line. X will be uniform in $[0, d]$. Finally, we will assume X and Θ to be independent.

An instant of reflection allows us to compute the desired probability, which we denote p:

$$p = P\left(X \le \frac{L}{2} \sin \Theta\right) + P\left(X + \frac{L}{2} \sin \Theta \ge d\right) \tag{123}$$

Under the assumptions we made, we will have:

$$P\left(X \le \frac{L}{2} \sin \Theta\right) = \int_{\{x \le \frac{L}{2} \sin(\theta)\}} dx d\theta p_X(x) p_\Theta(\theta)$$

$$= \frac{2}{\pi d} \int_0^{\frac{\pi}{2}} d\theta \int_0^{\frac{L}{2} \sin(\theta)} dx = \qquad (124)$$

$$= \frac{2}{\pi d} \int_0^{\frac{\pi}{2}} d\theta \frac{L}{2} \sin(\theta) = \frac{L}{\pi d}$$

Performing a similar calculation for $P\left(X + \frac{L}{2} \sin \Theta \ge d\right)$, we can conclude that:

$$p = \frac{2L}{\pi d} \qquad (125)$$

If we sample X and Θ we can thus estimate π.

5.2 We can model the situation as follows. Let Y is uniform in the square $[-1, 1] \times [-1, 1]$. Let $\mathscr{C}$ be the unit circle. Consider the probability law:

$$A \in \mathbb{R}^2 \to P(Y \in A \mid Y \in \mathscr{C}) \qquad (126)$$

which is the law we are sampling if we reject the realizations of Y that fall outside the circle.

We need to show that this is the law of a random variable, say X, uniform in $\mathscr{C}$. If $p_X(\mathbf{x})$ denotes the probability density of X, we will have:

$$\int_A d\mathbf{x} p_X(\mathbf{x}) = P(X \in A) = P(Y \in A \mid Y \in \mathscr{C}) = \frac{P(Y \in A \cap Y \in \mathscr{C})}{P(Y \in \mathscr{C})}$$

$$= \frac{4}{\pi} P(Y \in A \cap Y \in \mathscr{C}) = \frac{4}{\pi} \frac{\int_A d\mathbf{x} 1_\mathscr{C}(\mathbf{x})}{4} \qquad (127)$$

so that we have:

$$p_X(\mathbf{x}) = \frac{1}{\pi} 1_\mathscr{C}(\mathbf{x}) \qquad (128)$$

5.3 This exercise is a direct application of the transformation law of probability densities. In order to simplify the algebra, we can solve the problem in two-steps.

We start from $X = (X_1, X_2)$ uniform in the unit circle, and we first investigate the probability density for the polar coordinates $(R, \Theta) = \phi(X_1, X_2)$, where:

$$\phi(x_1, x_2) = \left(\sqrt{x_1^2 + x_2^2}, \arctan\left(\frac{x_2}{x_1}\right)\right) \qquad (129)$$

The inverse is naturally:

$$\phi^{-1}(r, \theta) = (r \cos \theta, r \sin \theta) \qquad (130)$$

and the jacobian is:

$$J_{\phi^{-1}}(r,\theta) = \begin{pmatrix} \cos\theta & -r\sin\theta \\ \sin\theta & r\cos\theta \end{pmatrix}, \quad \det\left(J_{\phi^{-1}}(r,\theta)\right) = r \tag{131}$$

So that:

$$p_{(R,\Theta)}(r,\theta) = p_X(\phi^{-1}(r,\theta)) \left|\det\left(J_{\phi^{-1}}(r,\theta)\right)\right| = \frac{r}{\pi} 1_{[0,1]}(r) \tag{132}$$

Now, we consider $Y = (Y_1, Y_2) = \psi(R, \Theta)$ where:

$$\psi(r,\theta) = \left(r\cos\theta \sqrt{\frac{-2\log r^2}{r^2}} , \; r\sin\theta \sqrt{\frac{-2\log r^2}{r^2}}\right) \tag{133}$$

It is easy tho check that:

$$\psi^{-1}(y_1, y_2) = \left(e^{-\frac{1}{4}(y_1^2+y_2^2)}, \; \arctan\left(\frac{y_2}{y_1}\right)\right) \tag{134}$$

and:

$$J_{\psi^{-1}}(y_1, y_2) = \begin{pmatrix} -\frac{y_1}{2} e^{-\frac{1}{4}(y_1^2+y_2^2)} & -\frac{y_2}{2} e^{-\frac{1}{4}(y_1^2+y_2^2)} \\ -\frac{y_2}{y_1^2} \frac{1}{1+\frac{y_2^2}{y_1^2}} & \frac{1}{y_1} \frac{1}{1+\frac{y_2^2}{y_1^2}} \end{pmatrix}$$

$$\det\left(J_{\psi^{-1}}(y_1, y_2)\right) = -\frac{1}{2} e^{-\frac{1}{4}(y_1^2+y_2^2)} \tag{135}$$

so that, finally:

$$p_Y(y_1, y_2) = p_{(R,\Theta)}\left(\psi^{-1}(y_1, y_2)\right) \left|\det\left(J_{\psi^{-1}}(y_1, y_2)\right)\right| = \frac{1}{2\pi} e^{-\frac{1}{2}(y_1^2+y_2^2)} \tag{136}$$

which is the desired result.

5.4 The random variable T_n represents the time instants when the n-th event happens. Under the assumptions of this exercise we can easily compute its characteristic function. As a preliminary step, we observe that, since ΔT_i is exponential, we have:

$$\phi_{\Delta T_i}(\theta) = \frac{\lambda}{\lambda - i\theta} \tag{137}$$

so that:

$$\phi_{T_n}(\theta) = \left(\frac{\lambda}{\lambda - i\theta}\right)^n \tag{138}$$

It is not difficult to show that this characteristic function corresponds to the probability density:

$$p_{T_n}(x) = \chi_{(0,+\infty)}(x) \frac{\lambda^n}{(n-1)!} x^{n-1} e^{-\lambda x} \tag{139}$$

which defines the **gamma law** $\Gamma(n, \lambda)$.

Now, the key observation is the following:

$$\{N_t \geq n\} = \{T_n \leq t\} \tag{140}$$

The interpretation is as follows: n events or more have happened by time t if an only if the n-th event happened no later than t. So, in order to solve the problem, we need to show that:

$$\sum_{i=n}^{+\infty} \frac{(\lambda t)^i}{i!} e^{-\lambda t} = \int_0^t dx \frac{\lambda^n}{(n-1)!} x^{n-1} e^{-\lambda x} \tag{141}$$

A simple way to verify this identity is to notice that, at $t = 0$, both sides vanish, so that we should only check the the derivatives with respect to t coincide. We have:

$$\frac{d}{dt} \left(\sum_{i=n}^{+\infty} \frac{(\lambda t)^i}{i!} e^{-\lambda t} \right) = \sum_{i=n}^{+\infty} \left\{ \frac{\lambda^i i (t)^{i-1}}{i!} e^{-\lambda t} - \lambda \frac{(\lambda t)^i}{i!} e^{-\lambda t} \right\}$$

$$= \frac{\lambda^n n (t)^{n-1}}{n!} e^{-\lambda t} - \lambda \frac{(\lambda t)^n}{n!} e^{-\lambda t}$$

$$+ \frac{\lambda^{n+1} (n+1) (t)^n}{(n+1)!} e^{-\lambda t} - \lambda \frac{(\lambda t)^{n+1}}{(n+1)!} e^{-\lambda t} + \frac{\lambda^{n+2} (n+2) (t)^{n+1}}{(n+2)!} e^{-\lambda t} + \ldots$$

$$= \frac{\lambda^n t^{n-1}}{(n-1)!} e^{-\lambda t}$$

$$\tag{142}$$

since all the other terms cancel. By inspection this result coincide with the derivative of the right hand side of (141).

5.5 This problem just requires evaluations of one-dimensional elementary integrals.

5.6 This problem just requires evaluations of one-dimensional elementary integrals.

Problems of Chap. 6

6.1 $X_t = -B_t$ is a continuous process, with $X_0 = -B_0 = 0$ and its increments $X_t - X_s$ are independent of the past because they are the negatives of the increments of the Brownian motion. The distribution of $X_t - X_s = -(B_t - B_s)$ is normal with mean 0 and variance $t - s$, by the parity of the normal distribution. Therefore, X_t is a Brownian motion.

The process $X_t = \frac{1}{\sqrt{u}} B_{ut}$ has continuous paths and $X_0 = 0$. The increment $X_t - X_s = \frac{1}{\sqrt{u}} (B_{ut} - B_{us})$ is independent of the history of B before us, which is exactly the history of X before s. The distribution of the increment $X_t - X_s$ is normal with mean 0 and variance $Var[\frac{1}{\sqrt{u}} (B_{ut} - B_{us})] = \frac{tu-ts}{u} = t - s$. Therefore, X_t is a Brownian motion.

6.2 We have:

$$m(t) = E\left[B_t + vt\right] = vt \tag{143}$$

Moreover:

$$C(t, s) = E\left[(B_t + vt)(B_s + vs)\right] = \min(t, s) + v^2 ts \tag{144}$$

Finally, the probability density is:

$$p(x, t) = \frac{1}{\sqrt{2\pi t}} \exp\left(-\frac{1}{2}\frac{(x - vt)^2}{2t}\right) \tag{145}$$

6.3 To compute $F_{R_t}(x) = P[R_t \leq x]$ we write $R_t^2 = B_{1,t}^2 + B_{2,t}^2$ and:

$$F_{R_t}(x) = P[R_t^2 \leq x^2] = P[(B_{1,t}, B_{2,t}) \in B_1(x)] \tag{146}$$

where $B_1(x)$ is the disk of radius x centered at 0. Then:

$$F_{R_t}(x) = \int_{B_1(x)} \frac{e^{-\frac{x_1^2 + x_2^2}{2t}}}{2\pi t} \tag{147}$$

moving to radial coordinates we find $F_{R_t}(x) = \frac{x}{t} e^{-\frac{x^2}{2t}}$. The mean value and variance are $\sqrt{\frac{\pi}{2}}$ and $2t$ respectively.

6.4 Let B_t be a brownian motion on the real axis. The stochastic process:

$$R_t = \begin{pmatrix} \cos(B_t) \\ \sin(B_t) \end{pmatrix} \tag{148}$$

is called brownian motion on the unit circle. Find the law of the random variables $\cos(B_t)$, $\sin(B_t)$ and $B_t \bmod 2\pi$. The first two random variables correspond to the projections of R_t onto the x and y axis, and the latter to the angle of the particle.
   The event $-1 \leq \sin(x) \leq y$, where $y \in [-1, 1]$, is realized if and only if $x \in \left[2k\pi - \frac{\pi}{2} - \arcsin(y), 2k\pi - \frac{\pi}{2} + \arcsin(y)\right]$, where k is an integer number.

$$p(\sin(B_t) \leq y) = \sum_{k=-\infty}^{\infty} p\left(2k\pi - \frac{\pi}{2} - \arcsin(y) \leq B_t \leq 2k\pi - \frac{\pi}{2} + \arcsin(y)\right) \tag{149}$$

the right member of this equation is a known quantity, since B_t is a normal random variable with mean 0 and variance t. The probability distribution of $\sin(B_t)$ is found differentiating $p(\sin(B_t) \leq y)$ with respect to y:

$$p(y) = \sum_{k=-\infty}^{\infty} \partial_y \arcsin(y) p_t\left(2k\pi - \frac{\pi}{2} + \arcsin(y)\right) + \partial_y \arcsin(y) p_t\left(2k\pi - \frac{\pi}{2} - \arcsin(y)\right) \tag{150}$$

where $p_t(x)$ is the probability distribution of B_t. This series can be summed exactly, the result being:

$$p_t(y) = \frac{1}{2\pi}\theta_3\left(\frac{\arcsin(y)}{2} - \frac{\pi}{4}, e^{-\frac{t}{2}}\right) + \frac{1}{2\pi}\theta_3\left(-\frac{\arcsin(y)}{2} - \frac{\pi}{4}, e^{-\frac{t}{2}}\right) \quad (151)$$

where $\theta_3(x, y)$ is Jacobi's theta function:

$$\theta_3(x, e^{-y}) = \sum_{n=-\infty}^{\infty} e^{-n^2 y} e^{i2n\pi x} \quad (152)$$

Therefore:

$$p_t(y) = \frac{1}{2\pi} \frac{1}{\sqrt{1-y^2}} \left(\theta_3\left(\frac{\arcsin(y)}{2} - \frac{\pi}{4}, e^{-\frac{t}{2}}\right) + \theta_3\left(-\frac{\arcsin(y)}{2} - \frac{\pi}{4}, e^{-\frac{t}{2}}\right)\right) \quad (153)$$

We can find the probability distribution of $\sin(B_t)$ at long time t using the following property:

$$\lim_{y\to\infty} \theta_3(x, e^{-y}) = 1 \quad (154)$$

We obtain:

$$p_t(y) = \frac{1}{\pi} \frac{1}{\sqrt{1-y^2}} \quad (155)$$

The event $0 \leq x \bmod 2\pi \leq \alpha$, where $\alpha \in [0, 2\pi]$, is realized if and only if $x \in [2k\pi, 2k\pi + \alpha]$ where k is an integer number. Therefore:

$$p(B_t \bmod 2\pi \leq \alpha) = \sum_{k=-\infty}^{\infty} p\left(2k\pi \leq B_t \leq 2k\pi + \alpha\right) \quad (156)$$

and the probability distribution of $B_t \bmod 2\pi$ is readily found differentiating this quantity with respect to α:

$$p_t(\alpha) = \sum_{k=-\infty}^{\infty} p_t\left(2k\pi + \alpha\right) = \frac{1}{2\pi}\theta_3\left(\frac{\alpha}{2}, e^{-\frac{t}{2}}\right) \quad (157)$$

for long time, $p_t(\alpha) \to \frac{1}{2\pi}$: in this sense, we can say that the particle spreads uniformly on the unit circle.

6.5 The brownian bridge is a gaussian process, being the difference of two gaussian processes. Moreover:

$$E[X_t] = 0$$

$$E[X_t X_s] = E[B_t B_s - s B_t B_1 - t B_1 B_s + ts B_1^2] = \min(t, s) - ts = \min(t, s) - \min(t, s)\max(t, s)$$

so that, if $s < t$, we have $E[X_t X_s] = s(1 - t)$.

6.6 This is a special case of the following situation:

$$X_t = \sum_{k=1}^{N} p_k \, B_{k,t} \tag{158}$$

where $\{B_{k,t}\}$ is a family of independent Brownian motions with increments independent from the past, and $\sum_{k=1}^{N} p_k^2 = 1$. We have:

$$X_{t+\Delta t} - X_t = \sum_{k=1}^{N} p_k \left(B_{k,t+\Delta t} - B_{k,t} \right) \tag{159}$$

which shows that the increments are independent from the past.
    We have also:

$$E[X_t] = 0 \tag{160}$$

and:

$$E[X_t X_s] = 0 = \sum_{k,l=1}^{N} p_k p_l \, E\left[B_{k,t} B_{l,s} \right] = \sum_{k=1}^{N} p_k^2 \, \min(t, s) = \min(t, s) \tag{161}$$

Finally, we have:

$$E\left[X_t B_{1,t} \right] = p_1 t \tag{162}$$

6.7 We can write:

$$p_Z(z) = \int_0^\infty dt \, p_Z(z|T = t) \, p_T(t) = \int_0^\infty dt \, p_{B_t}(z) \frac{e^{-\frac{t}{\tau}}}{\tau} = \int_0^\infty dt \, \frac{e^{-\frac{z^2}{2t}}}{\sqrt{2\pi t}} \frac{e^{-\frac{t}{\tau}}}{\tau} = \frac{e^{-\sqrt{2}\frac{|z|}{\sqrt{\tau}}}}{\sqrt{2\tau}}$$

which is the density of a Laplace random variable with mean 0 and variance τ.

Problems of Chap. 7

7.1 $E[X_t] = \int_0^t E[B_s] ds = 0$.

$$E[X_t^2] = \int_0^t ds \int_0^t ds' \, E[B_s B_{s'}'] = \int_0^t ds \int_0^t ds' \, \min(s, s') = t^3/3$$

Since $X_t \sim N(0, t^3/3)$, we have $X_T/T \sim N(0, T/3)$.

7.2 Let $t_i = i\frac{t}{n}$, $i = 0 \ldots n$ be a partition of the interval $[0, t]$. We know that:

$$Y_t^{(n)} = \sum_{i=0}^{n-1} B_{t_i} \left(B_{t_{i+1}} - B_{t_i} \right) \tag{163}$$

converges to X_t as $n \to \infty$. Since:

$$Y_t^{(n)} = \sum_{i=0}^{n-1} \frac{B_{t_{i+1}}^2 - B_{t_i}^2}{2} - \sum_{i=0}^{n-1} \frac{\left(B_{t_{i+1}} - B_{t_i} \right)^2}{2} \tag{164}$$

The first sum is readily computed observing that $\sum_{i=0}^{n-1} B_{t_{i+1}}^2 - B_{t_i}^2 = B_t^2$. Moreover, $B_{t_{i+1}} - B_{t_i} \sim N\left(0, \frac{t}{n}\right)$ and since the increments of the brownian motion are independent on the past, then $\sum_{i=0}^{n-1} \left(B_{t_{i+1}} - B_{t_i} \right)^2 \sim \frac{t}{n}\chi^2(n)$ This random variable has mean t and variance $2\frac{t^2}{n}$. Taking the $n \to \infty$ limit we find:

$$Y_t^{(n)} \to B_t^2 - t \tag{165}$$

Consider now the process $Y_t = f(B_t) = B_t^2$. The Itô formula leads to the following expression for dY_t:

$$dY_t = dt + 2B_t dB_t \tag{166}$$

whence, recalling the linearity of the stochastic differential, we immediately find:

$$B_t dB_t = d\left(\frac{B(t)^2 - t}{2} \right) \tag{167}$$

whose solution is:

$$X_t = \int_0^t B_s dB_s = \frac{B_t^2 - t}{2} \tag{168}$$

7.3 Consider the function $\Phi(t, x, y) = xy$. Due to the Itô formula:

$$d\Phi = \partial_t \Phi + \partial_x \Phi dX_t + \partial_y \Phi dY_t + \frac{1}{2} \left[\partial_{xx} \Phi g_1^2(X_t) + 2\partial_{xy} \Phi g_1(X_t)g_2(Y_t) + \partial_{yy} \Phi g_2^2(Y_t) \right] dt$$

explicit calculation of the derivatives of Φ yields:

$$d(X_t Y_t) = X_t dY_t + Y_t dX_t + g_1(X_t)g_2(Y_t)dt$$

that is, by definition:

$$X_T Y_T - X_0 Y_0 = \int_0^T X_t dY_t + \int_0^T Y_t dX_t + \int_0^T g_1(X_t)g_2(Y_t)dt$$

or:

$$\int_0^T X_t dY_t = [X_t Y_t]_0^T - \int_0^T Y_t dX_t - \int_0^T g_1(X_t)g_2(Y_t)dt$$

similar to the typical integration by parts formula, with a correction of the form $\int_0^T g_1(X_t)g_2(Y_t)dt$. In the special case $Y_t = h(t), dY_t = h'(t)dt$ we have:

$$\int_0^T h'(t) X_t \, dt = [h(t) X_t]_0^T - \int_0^T h(t) \, dX_t$$

if moreover $X_t = B_t$:

$$\int_0^T h'(t) B_t \, dt = h(T)B_T - \int_0^T h(t) \, dB_t$$

7.4 We apply the Itô formula:

$$dX_t = f(X_t)dt + g(X_t)dB_t \to d\Phi_t = \partial_t \Phi_t dt + \partial_x \Phi_t dX_t + \frac{1}{2}\partial_{xx}\Phi_t g^2(X_t)$$

to the functions $\Phi(x,t) = x^2$ and $\Phi(x,t) = \sin(t+x)$. We get:

$$d\Phi_t = 2B_t dB_t + dt \qquad d\Phi_t = \left(\Phi_{t+\frac{\pi}{2}} - \frac{\Phi_t}{2}\right)dt + \Phi_{t+\frac{\pi}{2}} dB_t$$

7.5 We consider the function $\Phi(x,t) = xF(-t)$. We use Itô's Lemma, together with $F'(t) = F(t)f(t)$ to get:

$$d\Phi_t = -F'(-t)X_t dt + F(-t)dX_t = F(-t)g(t)dB_t$$

or:

$$\Phi_t = \Phi_0 + \int_0^t F(-s)g(s)dB_t \to F(-t)X_t = x_0 \int_0^t F(-s)g(s)dB_t$$

$$X_t = x_0 F(t) + F(t)\int_0^t F(-s)g(s)dB_t$$

7.6 If f is piecewise constant:

$$f(t) = \sum_{i=1}^K c_k 1_{[t_{i-1},t_i)}(t), \quad t_0 = 0 \tag{169}$$

I_t is a linear combination of increments of the brownian motion:

$$I_t = c_1(B_{t_1} - B_0) + c_2(B_{t_2} - B_{t_1}) + \cdots + c_n(B_t - B_{t_{n-1}}), \quad n \le K \tag{170}$$

and it is thus normal, as we already know. Given another instant t', we have:

$$I_{t'} = c_1(B_{t_1} - B_0) + c_2(B_{t_2} - B_{t_1}) + \cdots + c_m(B_{t'} - B_{t_{m-1}}), \quad m \le K \quad (171)$$

where the time instants coincide with the ones in the expression for I_t. Let's evaluate:

$$E[I_t I_{t'}] = \sum_{i=1}^{n} \sum_{j=1}^{m} c_i c_j E[(B_{t_i} - B_{t_{i-1}})(B_{t_j} - B_{t_{j-1}})] \quad (172)$$

If i or j are larger that $\min(m, n)$, and whenever $i > j$ or $i < j$, the two random variables in the expectation are independent, so that the contribution to the sum vanishes. We thus conclude that:

$$E[I_t I_{t'}] = \sum_{i=1}^{\min(n,m)} c_i^2(t_{i+1} - t_i) = \int_0^{\min(t,t')} f^2(s)ds \quad (173)$$

This results holds also for any $f \in L^2(0, T)$, as can be shown by approximating f with piecewise constant functions.

Problems of Chap. 8

8.1 Let's consider first the process:

$$Y_t = x^{-1} + r \int_0^t \exp\left(\left(rK - \frac{\beta^2}{2}\right)s + \beta B_s\right)dt \quad (174)$$

we have:

$$dY_t = r \exp\left(\left(rK - \frac{\beta^2}{2}\right)t + \beta B_t\right)dt \quad (175)$$

We have thus:

$$X_t = \phi(t, B_t, Y_t), \quad \phi(t, x_1, x_2) = \frac{\exp\left(\left(rK - \frac{\beta^2}{2}\right)t + \beta x_1\right)}{x_2} \quad (176)$$

We can apply multidimensional Itô formula:

$$\begin{aligned}
dX_t &= \left(rK - \frac{\beta^2}{2}\right)X_t dt + \beta X_t dB_t \\
&\quad - \frac{1}{Y_t^2}\exp\left(\left(rK - \frac{\beta^2}{2}\right)t + \beta B_t\right)dY_t + \frac{1}{2}\beta^2 X_t dt \\
&= \left(rK - \frac{\beta^2}{2}\right)X_t dt + \beta X_t dB_t - rX_t^2 dt + \frac{1}{2}\beta^2 X_t dt \\
&= rX_t(K - X_t)dt + \beta X_t dB_t
\end{aligned} \quad (177)$$

8.2 We introduce the auxiliary process:

$$U_t = \exp(-At) \, X_t \tag{178}$$

We have:

$$dU_t = -A \exp(-At) \, X_t dt + \exp(-At) \, dX_t \tag{179}$$

Using the equation we have:

$$dU_t = -A \exp(-At) \, X_t dt + \exp(-At) \, (AX_t dt + \sigma d B_t) = \exp(-At) \, \sigma d B_t \tag{180}$$

which immediately implies:

$$U_t = U_0 + \int_0^t \exp(-As) \, \sigma d B_s \tag{181}$$

or:

$$X_t = X_0 \exp(-At) + \int_0^t \exp(-A(t-s)) \, \sigma d B_s \tag{182}$$

8.3 Consider $f = (f_1, f_2) : \mathbb{R} \to \mathbb{R}^2$, and the process $(f_1(B_t), f_2(B_t))$. We apply Itô formula to the two components:

$$
\begin{aligned}
df_{1,t} &= f_1^{(1)} d B_t + \frac{1}{2} f_1^{(2)} dt \\
df_{2,t} &= f_2^{(1)} d B_t + \frac{1}{2} f_2^{(2)} dt
\end{aligned}
\tag{183}
$$

where $f_1^{(k)}$ denotes the k-th derivative. Comparing the above Itô differential with the considered equation, we have the conditions:

$$f_1^{(2)} = -f_1, \quad f_1^{(1)} = -f_2, \quad f_2^{(2)} = -f_2^{(2)}, \quad f_2^{(1)} = f_1 \tag{184}$$

It is simple to check that:

$$f(B_t) = (\cos(B_t), \sin(B_t)) \tag{185}$$

is a solution.

8.4 This is a simple example of a Kolmogorov equation, the generator being given by:

$$\mathscr{L}_t = \frac{1}{2} \frac{d^2}{dx^2} + \frac{b-x}{1-t} \frac{d}{dx} \tag{186}$$

8.5 We consider the process (deterministic):

$$Y_t = (X_t, P_t) \tag{187}$$

satisfying the differential equation:

$$\begin{cases} dX_t = \frac{\partial \mathcal{H}}{\partial \mathbf{p}}(X_t, P_t)dt \\ dP_t = -\frac{\partial \mathcal{H}}{\partial \mathbf{x}}(X_t, P_t)dt \end{cases} \tag{188}$$

which can be seen as a special case of a SDE with zero diffusion and drift:

$$b = \left(\frac{\partial \mathcal{H}}{\partial \mathbf{p}}, -\frac{\partial \mathcal{H}}{\partial \mathbf{x}} \right) \tag{189}$$

The related Fokker-Planck equation, that is the Liouville equation:

$$\begin{aligned} \frac{\partial}{\partial t}\rho &= -\frac{\partial}{\partial y_i}(b_i \rho) = -\sum_{i=1}^{N} \frac{\partial \mathcal{H}}{\partial \mathbf{x}_i}\left(\frac{\partial \mathcal{H}}{\partial \mathbf{p}_i}\rho \right) - \sum_{i=1}^{N} \frac{\partial \mathcal{H}}{\partial \mathbf{p}_i}\left(-\frac{\partial \mathcal{H}}{\partial \mathbf{x}_i}\rho \right) \\ &= \left[\frac{\partial \mathcal{H}}{\partial \mathbf{x}_i}\frac{\partial}{\partial \mathbf{p}_i} - \frac{\partial \mathcal{H}}{\partial \mathbf{p}_i}\frac{\partial}{\partial \mathbf{x}_i} \right]\rho \end{aligned} \tag{190}$$

where we have used Schwarz theorem from classic analysis. Remembering the definition of the Poisson brackets, we have the celebrated Liouville equation:

$$\frac{\partial}{\partial t}\rho = -\{\rho, \mathcal{H}\} \tag{191}$$

8.6 This is a simple application of Itô formula to the process:

$$X_t = Y_t^2, \quad dY_t = -\beta Y_t + \frac{1}{2}dB_t \tag{192}$$

The details of the algebra are left to the reader.

8.7 This is a simple application of Itô formula to the process:

$$Y_t = 2\arcsin(\sqrt{X_t}) \tag{193}$$

where X_t is the Wright-Fisher process. After some algebra, it turns out that:

$$Y_t = 2\arcsin(\sqrt{X_t}) = B_t + a \tag{194}$$

where the constant a is obtained imposing the initial condition $X_0 = x_0$ almost everywhere.

Printed in the United States
By Bookmasters